SpringerBriefs in Food, Health, and Nutrition

Springer Briefs in Food, Health, and Nutrition present concise summaries of cutting edge research and practical applications across a wide range of topics related to the field of food science, including its impact and relationship to health and nutrition. Subjects include: Food Chemistry, including analytical methods; ingredient functionality; physic-chemical aspects; thermodynamics Food Microbiology, including food safety; fermentation; foodborne pathogens; detection methods Food Process Engineering, including unit operations; mass transfer; heating, chilling and freezing; thermal and non-thermal processing, new technologies Food Physics, including material science; rheology, chewing/mastication Food Policy And applications to: Sensory Science Packaging Food Qualtiy Product Development We are especially interested in how these areas impact or are related to health and nutrition. Featuring compact volumes of 50 to 125 pages, the series covers a range of content from professional to academic. Typical topics might include:

A timely report of state-of-the art analytical techniques
A bridge between new research results, as published in journal articles, and a contextual literature review
A snapshot of a hot or emerging topic
An in-depth case study
A presentation of core concepts that students must understand in order to make independent contributions

More information about this series at http://www.springer.com/series/10203

Anthony Keith Thompson • Suriyan Supapvanich
Jiraporn Sirison

Banana Ripening

Science and Technology

 Springer

Anthony Keith Thompson
King Mongkut's Institute
of Technology Ladkrabang
Bangkok, Thailand

Suriyan Supapvanich
King Mongkut's Institute
of Technology Ladkrabang
Bangkok, Thailand

Jiraporn Sirison
King Mongkut's Institute
of Technology Ladkrabang
Bangkok, Thailand

ISSN 2197-571X   ISSN 2197-5728   (electronic)
SpringerBriefs in Food, Health, and Nutrition
ISBN 978-3-030-27738-3   ISBN 978-3-030-27739-0   (eBook)
https://doi.org/10.1007/978-3-030-27739-0

This Springer imprint is published by the registered company Springer Nature Switzerland AG
The registered company address is: Gewerbestrasse 11, 6330 Cham, Switzerland

Preface

There have been many textbooks published dealing exclusively with *Musa* (bananas and plantains) perhaps starting with the one written by my former boss and mentor Professor C.W. Wardlaw (*Banana Diseases: Including Plantains and Abaca* 24 editions published between 1961 and 1974) to several more excellent books. Also, several chapters in various books have also dealt with bananas.

The objective of this book was therefore to deal exclusively with how bananas are ripened. However, the technology used is based on, and has implications with, a whole range of factors. These include the cultivar, growing conditions and how and at what maturity the fruit are harvested and handled. Also, various postharvest treatments can be applied to fruit and how these impact with the ripening process can have effects on the fruit quality.

The ripening process and the changes that occur are therefore considered in detail. There is a well-established successful technology that has evolved and is applied, particularly in more developed temperate countries where bananas are imported, but this technology essentially applies to a single genotype. However, most bananas are marketed locally in the country where they are grown, and often, different technologies are used in the ripening process, which have technical, economic and health implication. All these factors are discussed, and interacting implications are considered. Also, there is often insufficient and even contradictory information on how different ripening methods affect fruit quality including carbohydrates, proteins, phenolic, vitamins, acidity and other phytochemicals as well as texture, colour and flavour.

Banana sea trade began a little more than 100 years ago, but it has helped create great economic empires and sustained the livelihoods of millions of people in producer countries as well as in importing countries. The international banana trade has been very fast in employing new advancements in refrigerated transport and ripening technologies, since they can reduce transport costs and losses and increase the distance they can be successfully transported.

Anthony Keith Thompson
Suriyan Supapvanich
Jiraporn Sirison

Abbreviations

1-MCP	1-methylcyclopropene
AA	ascorbic acid
ABA	abscisic acid
ACC	1-aminocyclopropane-1-carboxylic acid
ACC oxidase	an enzyme involved in biosynthesis of ethylene
ACC synthase	an enzyme involved in biosynthesis of ethylene
ACC synthesis	the reaction catalysed by ACC synthase
ACPd	activated carbon impregnated with palladium
ATP	adenosine triphosphate
AVG	aminoethoxyvinylglycine
CA	controlled atmosphere
CMC	carboxymethyl cellulose
DFE	dietary folate equivalent
DPPH	1,1-diphenyl-2-picrylhydrazyl
Ethrel	ethephon (2-chloroethylphosphonic acid)
GA	gibberellins
GA_3	gibberellic acid
IAA	indole-3-acetic acid
IC	inclusion complex powder for slow ethylene release, ethylene-alpha-cyclodextrin
Imazalil	1-[allyloxy-2,4-dichlorophenethyl]imidazole, a fungicide
LDPE	low-density polyethylene film
LPE	lysophosphatidylethanolamine
MA	modified atmosphere
mRNA	messenger ribonucleic acid
PAL	phenylalanine ammonia-lyase
PE	polyethylene film
PG	polygalacturonase
PME	pectin methylesterase
POD	pyrogallol peroxidase
PPO	polyphenol oxidase

pVACs	provitamin A carotenoids
RAE	retinol activity equivalent
RH	relative humidity
RNA	ribonucleic acid
SAM	s-adenosyl-methionine
TA	total acidity; titratable acidity
TBZ	thiabendazole (2-thiazol-4-yl benzimidazole), a fungicide
ton	2240 pounds (lbs) = 1016 kg
tonne	1000 kg
TPC	total phenolic compounds
TFC	total flavonoid compounds
TSS	total soluble solids
ULO	ultralow oxygen

Contents

About the Authors

Jiraporn Sirison is Associate Dean and Assistant Professor in the Faculty of Agro-Industry, King Mongkut's Institute of Technology, Ladkrabang, Thailand. She studied at Maejo University and Mahidol University both in Thailand and Kyoto University in Japan. She is a Nutrition Chemist whose teaching and research are mainly in the area of the effects of processing on nutrients in food and the application of food colloids in food nutrition. She teaches nutrition to undergraduates and has published many research papers on these subjects. Also, she was brought up on her parents' banana farm in Thailand.

Suriyan Supapvanich is Assistant Professor in Postharvest Technology in the Faculty of Industrial Education and Technology, King Mongkut's Institute of Technology, Ladkrabang, Thailand. His field of specialization is postharvest biology and biochemistry of tropical fruit and vegetables. He studied at Kasetsart University and King Mongkut's University Thonburi, both in Thailand, and the University of Nottingham, UK, where he was awarded a PhD degree. He subsequently worked at the University of Natural Resources and Life Sciences in Austria and lectured in the Faculty Natural Resources and Agro-Industry at Kasetsart University. He has 19 years teaching experience in courses on postharvest technology of agricultural products and introduction of food science and technology to undergraduate students and advanced postharvest technology to MSc students. He has also published more than 40 papers in international journals and 2 book chapters. He is currently coordinating postharvest research projects with KMITL Prince of Chumphon Campus, KMUTT, Thammasat University and Anhui Agricultural University in China.

Anthony Keith Thompson is Visiting Professor at King Mongkut's Institute of Technology Ladkrabang in Thailand and was formerly Professor of Plant Science, University of Asmara, Eritrea; Professor of Postharvest Technology and Head of Department, Cranfield University, UK; Team Leader on an EU project on restructuring the banana industry in the Windward Islands; Principal Scientific Officer, Tropical Products Institute, London; Postharvest Expert for the UN in Sudan, Yemen and Korea for the Food and Agriculture Organization, Ghana and Sri Lanka for the International Trade Centre and Gambia for the World Bank; Advisor to the British, Jamaican and Colombian Governments in postharvest technology of fruit and vegetables; Research Fellow in Crop Science, University of the West Indies, Trinidad; Demonstrator in Biometrics, University of Leeds; as well as Consultant for various commercial and government organizations in many countries throughout the world. He has published over 100 scientific papers and many scientific textbooks.

Chapter 1
Introduction

Bananas grow everywhere in the tropics as well as in many sub-tropical countries. FAOStat (2019) estimated that the area of production of bananas in 2017 was about 5.6 million hectares producing some 114 million tonnes. There are many genotypes, most of which are ripened and eaten as fresh fruit. Others may be cooked before eating either unripe or ripe. For eating fresh they are harvested before they are ripe and then either allowed to ripen or treated in some way to initiate the ripening process. The most obvious change in bananas during ripening is their peel colour. This has been codified into what is called a ripening index or a colour index (Table 1.1; Fig. 1.1), which is commonly used in many countries for communication in the ripening and retail trades and among research and extension workers for all the genotypes.

About 15% of the world's banana production is for the export trade, and is based on 'Cavendish' (Heslop-Harrison and Schwarzacher 2007). International trade in bananas is crucial for the economies of many developing countries and bananas are popular throughout the world, but can only be grown successfully in the tropics and to a much lesser extent in the sub-tropics. Therefore, an enormous quantity of bananas is transported around the world every day, which contribute to climate change. Ripening centres and retail distribution represent approximately 10% of total emissions of greenhouse gases in the banana value chain, from which 75% correspond to energy consumption, 22% to distribution centres and 1% to ethylene production (Calberto et al. 2015). Overseas transport and primary production of bananas were the main contributors to the total greenhouse gas emissions including the consumer stage that resulted in a 34% rise in carbon footprint, mainly due to high wastage (Svanes and Aronsson 2013).

© The Author(s), under exclusive license to Springer Nature Switzerland AG 2019
A. K. Thompson et al., *Banana Ripening*, SpringerBriefs in Food, Health, and Nutrition, https://doi.org/10.1007/978-3-030-27739-0_1

Table 1.1 Changes in peel colour and starch and sugar content of pulp of 'Gros Michel' during ripening, based on data from United Fruit Company 1942. Source: Wardlaw (1961) with modifications

Ripening index	Peel colour	Total sugars (%)	Starch (%)
1	Green	0.1–2.0	19.5–21.5
2	Green – trace of yellow	2.0–5.0	16.5–19.5
3	More green than yellow	3.5–7.0	14.5–18.0
4	More yellow than green	6.0–12.0	9.0–15.0
5	Green tips	10.0–18.0	2.5–10.5
6	All yellow	16.5–19.5	1.0–4.0
7	Yellow flecked with brown	17.5–19.0	1.0–2.5
8	Yellow with large brown spots	18.5–19.0	1.0–1.5

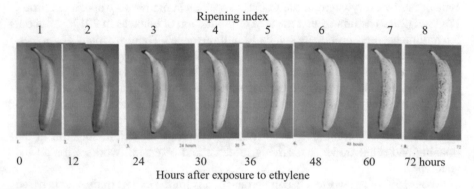

Ripening index

1   2   3   4   5   6   7   8

0   12   24   30   36   48   60   72 hours

Hours after exposure to ethylene

Fig. 1.1 Time lapse photographs of a single 'Valery' banana after 24 h exposure to 1000 µL L^{-1} ethylene and then kept at 20 °C and 90% RH

Musa Taxonomy

Bananas belong to the family Musaceae, genus *Musa*. Christelová et al. (2011) reported that *Musa* was divided into the following sections: *Ingentimusa* with a chromosome number $2n = 14$, *Australimusa* has been merged with *Callimusa* and has a chromosome number $2n = 20$ (*Musa beccarii*, which is part of *Callimusa* section, has 18 chromosomes) and *Eumusa* and *Rhodochlamys* with chromosome numbers of $2n = 22$. Fe'i belong to *Callimusa* and are easily recognized by their erect bunches.

Carl Linnaeus in 1783 gave the name *M. sapientium* L. to bananas that are eaten fresh and *M. paradisiaca* L. for those that are cooked before eating. Many authors still use the Linnaean classification and many other names have been used, but Cheesman (1947–1949) defined *M. balbisiana* Colla and *M. acuminata* Colla as the basis for almost all cultivated bananas and plantains. Simmonds and Shepherd

(1955) and Stover and Simmonds (1987) confirmed this and reported that many desert varieties are derived from *M. acuminata*, some being diploid and a few being tetraploid but most being triploid ($2n = 3x = 33$). *M. balbisiana* has also contributed to the origin of desert bananas and plantains by hybridisation with *M. acuminata*. For example, 'Pelipita' (*Musa* ABB) has 8 *M. acuminata* chromosomes and 25 *M. balbisiana* chromosomes. Simmonds and Shepherd (1955) recommended that in place of the species name, an A genome for *M. acuminate* and a B genome *M. balbisiana* should be used showing the origin and contribution of the two species. So, a triploid genotype whose origin is *M. acuminate*, e.g. 'Giant Cavendish' and 'Gros Michel' would be referred to as *Musa* AAA. A triploid that has one third *M. bulbisiana* and two thirds *M. acuminate* would be referred to as *Musa* AAB. They also recommended a subgroup within a species for example *Musa* AAA (Cavendish subgroup) 'Robusta', *Musa* AA 'Pisang Mas' and *Musa* ABBB 'Kluai Teparod'. The centre of origin is thought to be South-East Asia, where they occur from India to Polynesia (Simmonds 1962). The centre of diversity has been placed in Malaysia or Indonesia (Daniells et al. 2001), although considerable diversity is known throughout the range. The centre of origin of *M. acuminata* is thought to have been in Malaysia or Indonesia (Simmonds 1962) and *M. balbisiana* in India, Myanmar, Thailand and Philippines (Daniells et al. 2001). The origin of Fe'i is thought to be the Pacific Islands probably derive mainly from *Musa maclayi* F.J.H. von Mueller *ex* N.N. Miklouho-Maclay or *Musa* × *troglodytarum* L. They can be classified, for example, as *Musa* (Fe'i Group) 'Utafun'.

A taxonomic scoring method is used to classify the edible bananas and to provide evidence on their evolution. Edible diploid forms of *M. muminata* are thought to be the primary source of the whole group to which another species, *M. balbisiana*, has contributed by hybridization. Thus, there exist diploid and triploid edible forms of *M. muminata* and diploid, triploid and tetraploid hybrid types of genetic constitutions that vary according to their history. There is a slight possibility that a third wild species has contributed to the origins of a small group of triploid hybrid types. Triploidy was probably established under human selection for vigour and fruit size; tetraploidy is rare. Indo-Malaysia and Malaysia are probably the primary centre of origin of the group. The species *M. paradisiaca* and *M. sapientum*, named by Linnaeus to identifiable edible varieties, are both shown to be of hybrid origin (Simmonds and Shepherd 1955; Deepthi 2016).

Almost all genotypes used commercially in international trade have arisen from mutations of *Musa* AAA ('Cavendish' subgroup) selected in the field. These include 'Grand Naine', 'Lacatan', 'Poyo', 'Robusta', 'Valery' and 'Williams', which are largely distinguished from each other by the height of the pseudostem. In addition, 'Dwarf Cavendish' is tolerant to a wide range of climates including cool conditions. 'Grand Naine' responds well to optimum growing conditions, but does not grow or yield well in sub-optimum conditions including soils with low fertility or insufficient rainfall or irrigation. The 'Sucrier', 'Pisang Mas', 'Honey' genotypes are very sweet and have small fruit, thin skins, yellowish flesh and small bunches (up to about 13 kg). 'Gros Michel' was the fruit that dominated the international trade from its beginnings in the mid-nineteenth century until the late 1940s. It has long

been grown in South-East Asia and Sri Lanka. Jean Francois Pouyat, a French bota-
nist and chemist who settled in Jamaica in 1820 probably introduced it to the
Caribbean region. He brought the fruit from Martinique to his coffee estate in
Jamaica where it was originally called 'Banana Pouyat' then later this became the
'Martinique Banana' and finally 'Gros Michel' and 'Big Mike' in USA and 'Hom
Thong' in Thailand. The Agricultural Society in Jamaica awarded Pouyat a dou-
bloon for introducing such a valuable genotype (Von Loesecke 1949; Davies 1990).
It is a tall heavy bearing genotype, but is susceptible to Panama disease (caused by
the fungus *Fusarium oxysporum* f. sp. *cubense*). It was reported as early as 1928 in
Trinidad that there was concern about the susceptibility of 'Gros Michel' to Panama
disease (Wardlaw 1961). Three races of Panama disease have been described that
can affect bananas. Race 1 caused the epidemics on 'Gros Michel' and also affects
some other varieties and tetraploids. Race 2 affects cooking bananas e.g. 'Bluggoe'
and some tetraploids while Race 4 affects race 1 and race 2 susceptible varieties as
well as the 'Cavendish' varieties. Race 4 has been shown to consist of a sub-tropical
and a tropical race. The tropical race 4 can attack unstressed plants while sub-
tropical race 4 usually only attacks when plants are in a stressed condition. 'Namwa'
(AAB) was reported to be very resistant to *Fusarium oxysporum* f. sp. *cubense*,
tropical race 4. Infection incidence were less than 1% while susceptible varieties
like 'Lakatan' and 'Cavendish' sustained more than 90% in the first crop. 'Namwa'
was also observed resistant to Banana Bunchy Top Virus (BBTV) and Black
Sigatoka (Molina 2019).

Musa Breeding Programmes

Breeding programmes have developed tetraploid bananas and plantains, starting in
1922 at the Imperial College of Tropical Agriculture in Trinidad. This programme
was subsequently transferred to establish the banana breeding programme at Bodles
in Jamaica (producing, among others, 'Bodles Altafort' which is *Musa* AAAA).
Other breeding programmes include the Centro Nacional de Pesquisa de Mandioca
e Fruticultura Tropical and Empresa Brasileira de Pesquisa Agropecuária
(EMBRAPA) that were established in Brazil and the Fundacion Hondureña de
Investigacion Agricola (FHIA) that was originally established at a research station
of the United Fruit Company in Honduras. These programmes have largely concentrated in breeding tetraploids. 'FHIA-21' (*Musa* AAAB), with cooking banana with
plantain in its pedigree, has been reported to be resistant to Black Sigatoka disease.
'FHIA-01', also called 'Goldfinger' (*Musa* AAAB), is a desert banana that has been
bred to be high yielding and resistant to Black Sigatoka, however this resistance has
been questioned. There is also concern about the tetraploids containing *M. balbisiana* because the Banana Streak Virus (BSV) is integrated in the B genome and may
be activated, especially if they are exposed to stress or are propagated by tissue
culture (Jones 2009). The International Institute of Tropical Agriculture (IITA) in
Nigeria and Kenya have programmes involving crossing *Musa* AA with *Musa* AAB

to generate triploid and tetraploids (*Musa* AAAB) with improved disease resistance and agronomic characters. Several cultivars have been released by ITA that are either triploid or tetraploid and are given numbers following the initials, PITA for plantains, for example PITA 14, and BITA for cooking bananas, for example BITA 3. Centre Africain de Recherches sur Bananiers et Plantains (CARBAP) in West Africa and Centre de cooperation Internationale en Recherche Agronomique pour le Développement (CIRAD) in the Guadeloupe also have banana breeding projects.

Genotypes in International Trade

Currently almost all varieties used commercially in international trade have arisen from mutations of *Musa* AAA 'Cavendish' subgroup (Fig. 1.2) selected in the field including 'Grand Naine', 'Lacatan', 'Poyo', 'Robusta', 'Valery' and 'Williams'. From their origin in South-East Asia some authorities report that bananas were introduced into Africa in the early middle ages, and in 1402 Portuguese colonists took banana plants from Guinea to the Canary Islands. They were thought to have been taken to the Americas by Portuguese explorers in the late fifteenth century. In 1826 Charles Telfair obtained 'Cavendish' plants from southern China and took them to Mauritius and in 1829 some plants were sent from there to England. In 1834, the 6th Duke of Devonshire wrote to the Chaplain at Alton Towers: "My Dear Sir, A thousand thanks for the banana, it arrived quite safe and I am delighted to have an opportunity of seeing that most beautiful and curious fruit. It is the

Fig. 1.2 'Giant Cavendish' bunch ready for harvesting in St Lucia

admiration of everybody and has been feasted upon at dinner today according to the directions". The Duke eventually obtained plants at Chatsworth House, which is the seat of the Dukes of Devonshire. Their family name is 'Cavendish' and the Duke named it *Musa cavendishii*. Joseph Paxton was their head gardener who cultivated the bananas and got them to produce fruit in a conservatory at Chatsworth. A few years later the Duke supplied two cases of plants to a missionary named John Williams, destined for Samoa. Only one *M. cavendishii* plant survived the journey and that single specimen was the forebear of bananas that flourish in Samoa and other South Sea Islands. Probably in 1853 some plants were received from Chatsworth and although they were in poor condition some survived and were distributed among the islands. Also, the first banana plant from Chatsworth may also have been the origin of stock introduced to the Canaries in 1855 (Anonymous 2012). Apparently, suckers were also sent to the West Indies about the same time.

Some Common Genotypes

As indicate above the botanical names for different bananas are based on the publications of Cheesman (1947–1949) and subsequently those of Simmonds and Shepherd (1955) and Stover and Simmonds (1987). A cultivar is a plant that has been produced in cultivation and a variety is a naturally occurring form of plant, both are within a species. Therefore, most of the names used for bananas could be called varieties since they have occurred naturally through selections of somatic mutations from a cultivated population. However, a few, particularly the tetraploids, are cultivars since they have been specifically bred for certain desirable characteristics.

There are many local names used for different types of banana therefore deciding which names are the same and which names are different is difficult and at best arbitrary. Some botanical names are different depending on the references, for example 'Bluggoe' is given as *Musa* AAB (Christelová et al. 2017) and *Musa* ABB (Porcher 1998), but for simplicity the one given by INIBAP (Daniells et al. 2001) will be used. Also, transliterations between languages can cause confusion. In some cases, different botanical names are given for fruit with the same common name. Therefore, the following list of banana genotypes, mostly referred to in this book, is by no means definitive, just an indication and an attempt to reduce confusion. With the susceptibility of bananas and plantains to somatic mutation, many small-scale growers will have their own often exclusive selection that may be unique. Here they are listed under common names to provide a reference. For a much more detailed and comprehensive description, refer to Stover and Simmonds (1987), Daniells et al. (2001) and the Musa Germplasm Information System.

'Apem' *Musa* AAB (plantain subgroup).
'Apple banana' *Musa* AA 'Manzano' has a sub-acid flavour and should be allowed to ripen fully before eating.

'Amritsagar' *Musa* AAA is a good table variety in Assam with medium sized plant, good size fruit and medium thick rind, and the ripe banana develops a bright yellow colour (Goswami and Handique 2013).

'Bananito' *Musa* AA 'Kluai Namwa' 'Lady Finger' (Fig. 1.3).

'Bluggoe' *Musa* ABB 'Kluai Som', 'Horse Plantain' tends to be highly susceptible to *Fusarium oxysporum* f. sp. *cubense* race 1, 2 and 4 (Jones 2000).

'Bodles Altafort' *Musa* AAAA developed from the banana breeding programme at Bodles in Jamaica by Ken Shepherd in the 1960s and was a cross between 'Gros Michel' (AAA) and 'Pisang Lilin' (AA).

'Cavendish' *Musa* AAA. has been the main group of varieties in international trade since the late 1940s and selections include 'Grand Naine', 'Lacatan', 'Poyo', 'Robusta', 'Valery' and 'Williams', which are largely distinguished from each other by the height of the pseudostem. The tallest is 'Lacatan' (up to 5.5 m) followed by 'Robusta' and 'Giant Cavendish' (up to 5 m) and the smallest is the 'Dwarf Cavendish' (1.2–2.1 m). 'Grand Naine' responds well to optimum growing conditions, but does not grow or yield well in sub-optimum conditions including soils with low fertility or insufficient rainfall or irrigation.

'Cheni Champa' *Musa* AAB is one of the hardiest medium tall bananas in Assam, resistant to Fusarium wilt and fairly resistant to bunchy top disease with small size fruits, thin peel, creamy pulp and sub-acid taste (Goswami and Handique 2013).

'Giant Cavendish' *Musa* AAA is synonymous with 'Cavendish'.

'Dominico' *Musa* AAB (plantain subgroup).

'Dominico Harton' *Musa* AAB (plantain subgroup).

'Dwarf Brazilian' *Musa* AAB.

'Dwarf Cavendish' *Musa* AAA (Cavendish subgroup). 'Dwarf Cavendish' is tolerant to a wide range of climates including cool conditions.

'FHIA-01' *Musa* AAAB 'Goldfinger', is a desert banana that can also be used in cooking when green. It was bred to be high yielding and resistant to Black

Fig. 1.3 *Musa* AA on sale in UK supermarkets in 2018 as 'Bananitos'

Sigatoka. Mean bunch weight 34.1 lbs, taste 3.2 and acceptability 3.7 (Nelson and Javier 2007).

'FHIA-02' *Musa* AAAA 'Mona Lisa' is sweet and similar to Cavendish. Mean bunch weight 23.7 lbs, taste 3.0 and acceptability 3.6 (Nelson and Javier 2007).

'FHIA-03' *Musa* AABB 'Sweetheart' a cooking banana that can also be used for dessert. Mean bunch weight 36.4 lbs, taste 3.0 and acceptability 3.6 (Nelson and Javier 2007).

'FHIA-17' *Musa* AAAA a 'Gros Michel' type dessert banana that can also be cooked. Mean bunch weight 27.4 lbs, taste 3.3 and acceptability 4.2 (Nelson and Javier 2007).

'FHIA-18' *Musa* AAAB a sweet acid apple flavoured dessert banana. Mean bunch weight 27.6 lbs, taste 2.9 and acceptability 3.8 (Nelson and Javier 2007).

'FHIA-21' *Musa* AAAB is a French plantain type.

'FHIA-23' *Musa* AAAA dessert banana. Mean bunch weight 26.7 lbs, taste 3.0 and acceptability 3.7 (Nelson and Javier 2007).

'Grand Naine' *Musa* AAA (Cavendish subgroup). It responds well to optimum growing conditions, but does not grow or yield well in sub-optimum conditions including soils with low fertility or insufficient rainfall or irrigation. Mean bunch weight 18.3 lbs, taste 3.2 and acceptability 3.1 (Nelson and Javier 2007).

'Gros Michel' *Musa* AAA dominated the international trade from its beginnings in the mid-nineteenth century until the late 1940s. It is a tall heavy bearing genotype, but is susceptible to Panama disease (caused by the fungus *Fusarium oxysporum* f. sp. *cubense*). It is still grown commercially, for example in parts of Colombia and Thailand where it is mainly called 'Kluai Hom Thong'.

'Guineo' *Musa* AAA.

'Guineo Negro' *Musa* AAA.

'Hari chhal' (*Musa acuminata*) grown commercially in Sindh Province in Pakistan and in India.

'Harton' *Musa* AAB (plantain subgroup).

'Highgate' *Musa* AAA semi-dwarf mutant of 'Gros Michel'.

'Honey' *Musa* AA.

'Horn' *Musa* AAB (plantain subgroup).

'Khai' *Musa* AA (Fig. 1.4).

'Karpooravalli' 'Karpuravalli' *Musa* ABB.

'Khuai Hom Khiew Korm' synonymous with 'Dwarf Cavendish' *Musa* AAA (Fig. 1.5).

'Kluai Hom Khiew' *Musa* AAA (Cavendish subgroup) synonymous with 'Pisang Masak Hijau'.

'Kluai Hom Taiwan' *Musa* AAA synonymous with 'Gros Michel'.

'Kluai Hom Thong' *Musa* AAA synonymous with 'Gros Michel'.

'Kluai Hom' *Musa* AAA (Cavendish subgroup).

'Kluai Hug Mook' *Musa* AA synonymous with 'Silver Bluggoe', 'Kluai Som'.

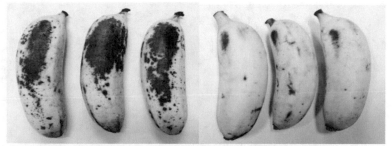

after storage for 8 days at 25°C   after storage for 16 days at 13°C

Fig. 1.4 'Khai' fruit at harvest and after storage at 13 or 25 °C

Fig. 1.5 'Hom Khiew' on sale in Thailand in 2019

'Kluai Khai' *Musa* AA synonymous with: 'Kluai kha' 'Pisang Mas', 'Sunny Bunch', 'Golden', 'Sucrier'.
'Kluai Leb Mua Nang' *Musa* AA (Fig. 1.6).

'Kluai Nam Wah' *Musa* AA synonymous with: 'Pisang Awak', 'Kluai Hin' and 'Saba'.
'Kluai Nark' *Musa* AAA (Cavendish subgroup) synonymous with 'Red banana' 'Lacatan'.
'Kluai Teparod' *Musa* ABBB.
'Lady Finger' *Musa* AA (Fig. 1.7).

'Latundan' *Musa* ABB.
'Malbhog' *Musa* AAB synonymous with: 'Amrithapani', 'Digjowa', 'Dudhsagar', 'Honda', 'Kozhikodu', 'Latundan', 'Madhuranga', 'Manzana', 'Nanjankode Rasabale', 'Pisang Rasthali' 'Silk', 'Silk Fig' 'Saapkal', 'Sabri', 'Suvandal' and 'Thozhuvan'. It is a popular table banana in Assam with a sweet aroma, taste and long postharvest life (Goswami and Handique 2013).

Colour index 1   Colour index 7

Fig. 1.6 'Kluai Leb Mua Nang' in Thailand in 2019 at harvest and after ripening in ambient conditions (about 30 °C)

Fig. 1.7 Golden, Lady
Finger on sale in Thailand
in 2019

'Namwa' *Musa* AAB synonymous with: 'Pisang Awak', which is a popular banana
  in Thailand and Malaysia. Flavours were similar to 'Cavendish' except that they
  were stronger (Renoo Yenket, nd).
'Nanica' *Musa* AAA (Cavendish subgroup).
'Ney Poovan' *Musa* AB 'Safet Velchi', 'Chini Champa', 'Ranel', 'Kisubi', 'Apple',
  'Farine France', 'Lady's Finger' has a sweet, sub-acid flavoured fruit.
'Orishele' *Musa* AAB.
'Pacovan' *Musa* AAB, subgroup Prata. is mostly cultivated in Northeast Brazilian
  but is also grown in India, Australia and the Western Pacific islands, where it is
  known as 'Pachanadan' or 'Pacha Naadan', 'Improved Lady's Finger' and
  'Lady's finger' respectively (Ploetz et al. 2007).
'Pei Chiao' *Musa* AAA (Cavendish subgroup) introduced to Taiwan from southern
  China in the eighteenth century.
'Petite Naine' *Musa* AAA (Cavendish subgroup) synonyms include: 'Extra-dwarf',
  'Kiri Tia Mwin' and 'Dwarf Cavendish'.

'Pisang Ambon Lumut' *Musa* AAA.

'Pisang Awak' *Musa* AAB.

'Pisang Mas' *Musa* AA synonyms include: 'Amas', 'Sucrier', 'Kluai Khai', 'Figue Sucrée.

'Pisang Lilin' *Musa* AA synonyms include: Kluai Thong Ki Maew, Kaveri banana. Pisang ekor kuda, Pisang empat puluh, Pisang lemak manis, Pisang lemak manis terenganu, Pisang lidi, Pisang mas sagura, Pisang muli, Kluai Leb Mu Nang, Kluai Thong Kab Dam, Kluai tong kee maew Kluai thong khi meew.

'Pisang Ustrali' *Musa* AAAA synonyms with Papua New Guinea banana.

'PITA 3' *Musa* AAAB.

'PITA 314' *Musa* AAAB.

'PITA 24' plantain *Musa* AAB.

'Pouyat' *Musa* AAA 'Gros Michel' is named after Jean Francois Pouyat, a French botanist and chemist who settled in Martinique.

'Poyo' *Musa* AAA.

'Prata'*Musa* AAB. Approximately 60% of the harvested area of bananas in Brazil are 'Prata', 'Prata Anã' or 'Pacovan' (Embrapa 2014).

'Prata Anã' *Musa* AAB synonyms with: 'Lady's finger' and 'Santa Catarina Prata' in Hawai'i.

'Rasabale' *Musa* AAB (Silk subgroup).

'Red Decca' *Musa* AAA (Cavendish subgroup).

'Red Banana' *Musa* AAA (Cavendish subgroup).

'Robusta' *Musa* AAA (Cavendish subgroup).

'Saba' *Musa* BBB.

'Santa Catarina Prata' *Musa* AAA.

'Sucrier' *Musa* AA. 'Pisang Mas' and 'Honey' are very sweet and have small fruit, thin skins, yellowish flesh and small bunches (up to about 13 kg). Plants are up to 3.5 m high, prefer light shade and are not well adapted to cooler temperatures.

'Sugar' *Musa* AAB.

'Valery' *Musa* AAA (Cavendish subgroup).

'Williams' *Musa* AAA (Cavendish subgroup).

Where bunch flavour and acceptability are given above, these are based on the scale of Nelson and Javier (2007) where: 5 = excellent, 4 = very good, 3 = good, 2 = fair, 1 = poor.

Chapter 2
Preharvest Effects

The conditions and environment in which a crop is grown can affect its postharvest life including ripening of fruit. Chillet and de Lapeyre de Bellaire (2002) found a weak correlation between the manganese concentration and wound ethylene production in lowland bananas grown in the West Indies. They also showed that in the wet season lowland fruit were more fragile and produced more wound ethylene than highland fruit. In Thailand 'Williams' and 'Grand Naine' bananas grown under low chemical production systems tended to have higher levels of sugars and acids but were softer and there was some indication that they contained less starch than those from a conventional production system (Ambuko et al. 2006).

Fertilizers

No direct information could be found on the effects of fertilizer application to the growing plant on the subsequent behaviour during ripening of bananas. However, Ramesh Kumar and Kumar (2007) assessed potash foliar sprays on 'Neypoovan' bananas and found their postharvest green life was 4.5, 4.8, 5.2 and 5.3 days as a result of foliar sprays with 0, 0.5, 1.0 and 1.5% P, respectively. Also, TSS, total sugars and sugar acid ratio increased and weight loss and % acidity decreased as K foliar sprays were increased. Patel et al. (2017) found marked improvement in fruit quality of 'Grand Naine' when RDK (a concentrated N, P, S and Mg fertilizer) was applied to the plants. The quality factors they were assessing were: pulp to peel ratio, TSS, TA, ascorbic acid, reducing sugars, total sugars, sugar acid ratio and shelf-life.

In a long-term study of the effects of N, P and K fertilizer levels on postharvest respiration of apple, Letham (1969) found that fruit that had had the highest P concentration had the lowest respiration rate while those with the lowest P concentration had the highest respiration rate. Apples grown with low P initiated the climacteric sooner than those grown with high P, indicating that lower levels of tissue P were sufficient to alter fruit developmental physiology. Lin and Ehret (1991) showed that the shelf-life of greenhouse grown cucumbers could be improved by increasing the concentrations of N, P, K, Ca, S and Mg in the nutrient solution. Cucumbers were also grown in a greenhouse under low and high P fertilizer regimes by Knowles et al. (2001). The respiration rate of low P fruits was 21% higher than that of high P fruits and they began the climacteric rise about 40 h after harvest, reached a maximum at 72 h and declined to pre-climacteric levels by 90 h. The difference in respiration rate between low and high P fruits was up to 57% during the climacteric. The climacteric was different to the low P fruits and was not associated with an increase in fruit ethylene concentration or ripening.

Organic Production

In a case study of the postharvest qualities of conventionally versus organically grown bananas in the Dominican Republic, Caussiol and Joyce (2004) found that significant differences ($p < 0.05$) in measured quality attributes between conventionally and organically grown fruit were "few and marginal". Moreover, any differences were inconsistent across harvest-times and during shelf-life. Thus, organically and conventionally grown bananas had almost identical qualities. Sensory comparison confirmed that there was no flavour difference. Nyanjage et al. (2000) found that organically grown 'Robusta' bananas ripened faster at 22–25 °C than non-organically grown bananas as measured by peel colour change, but ripe fruit had similar TSS levels from both production systems. The peel of non-organic fruits had higher N and lower P contents than organic fruits. In a survey of fruit quality of Philippine bananas from non-chemical production, the problems highlighted all related to management practices and none to the effects of organic production on postharvest aspects (Alvindia et al. 2000).

Light and Day Length

Banana is a day neutral crop and no information could be found on the effects of light intensity or day length and banana ripening, although direct exposure to the sun can damage the fruit (Fig. 2.1). Work has been carried out on other fruit. For example, Woolf et al. (2000) showed that during ripening of avocados at 20 °C, fruit that had been exposed to direct sunlight showed a delay of 2–5 days in their ethylene peak compared with fruits that had been grown in the shade. Aborisade and Ayibiowu (2010) studied the effects of day length on ripening of 'Roma' and 'Beske' tomatoes. They were harvested at the mature-green, breaker, turning or pink stages and then stored at 28 °C either in alternating light, 12 h light and 12 h in

Fig. 2.1 Sun damage on
bananas in Eritrea

complete darkness. Ripening progressed in complete darkness more quickly than in the alternating light and dark in fruit at the breaker stage. However, the difference was more pronounced in 'Roma' than in 'Beske'. Fruit initially at the turning stage did not show any significant effect of photoperiod in 'Roma' but did by day 6 in 'Beske'. Generally, photoperiod had more effect at the mature-green and breaker stages than at the turning and pink stages of ripening.

Disease

Generally, if a crop has suffered an infection during development its storage or marketable life may be adversely affected. Bananas may ripen prematurely or abnormally after harvest because of leaf infections by fungi during growth, which caused stress and therefore shortened their storage life. This can be observed on the crop before harvesting or it may only be observed as a physiological disorder postharvest. Fungicide applications in the field to control both leaf spot diseases, *Mycosphaerella musicola* and *M. fijiensis*, were shown to reduce premature ripening (Thompson and Burden 1995). In addition, yield was reduced in areas where plants were infected with leaf spot, as it can become necessary to harvest fruit at a lower grade (i.e. younger) in order to lessen the chances of premature ripening of the fruit in transit (Stover 1972; Stover and Simmonds 1987). Black Sigatoka is caused by infection with *M. fijiensis*, which is endemic to most banana exporting countries. It does not infect the fruit, but infection can cause damage to leaves or even kill them. This reduces the photosynthetic area of the plant that can lead to reduction in yield. This leaf damage in turn causes stress to the plants and a reduction in the pre-climacteric life of fruit harvested from infected plants. Bananas harvested from Black Sigatoka infected plants may behave as though they were physiologically 1–2 weeks older than those of non-infected plants of the same age. In trials reported by Turner (1997), reducing the number of leaves from 12 to 7 during fruit growth did not affect bunch weight, but reduced pre-climacteric life of fruit by 6 days. With even fewer leaves there was no further reduction in pre-climacteric life, but bunch weight was reduced by 8%. Similar results were reported by Ramsey et al. (1990) who found that plants with less than five of their leaves not greatly affected by Black Sigatoka at harvest produced smaller bunches. All bunches from plants with fewer than 4 leaves were "field-ripe", that is suffering from Pulpa Crema. Pulpa Crema is where bananas are ripe but the peel remains green. This is because the bananas are initiated to ripen on the plant, but because the ambient temperature is above 25 °C the pulp ripens but the chlorophyll in the skin is not fully broken down. This effect occurs in 'Cavendish' but not in plantains (Fig. 2.2) (Seymour et al. 1987) and many other *Musa* genotypes including 'Kluai Hom Thong'.

Since the introduction of the shipment of bananas as hands packed in fibreboard boxes the principal postharvest disease problem has been Crown Rot. When the export trade began bananas were transported as bunches. It was not until bunches were cut into hands or clusters for packing into fibreboard boxes that Crown Rot became a problem that needed treating. For local marketing in many producing countries, a portion of the stem is retained still attached to the crown (Figs. 1.5 and 1.6), which appears to prevent Crown Rot. Crown Rot develops from the infection of the cut surface of the crown, where it has been cut from the fruit stalk, by fungi. If not treated these infections develop during transport and ripening in the importing country and cause decay of the crown tissue, which may spread into the fruit pedicel or even the fruit itself during ripening and marketing. Several different fungi have been found to be associated with Crown Rot (Griffee and Pinegar 1974; Griffee and Burden 1976), the most common being the banana anthracnose fungus *Colletotrichum*

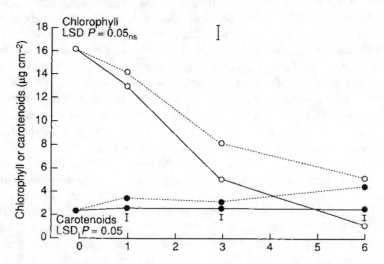

Fig. 2.2 Changes in the pigment level of 'Cavendish' bananas during ripening at either 20 °C (.....) or 35 °C (___) for 6 days. The fruit were exposed to 1000 μL L^{-1} before being ripened in air. (Source: Seymour 1985)

musae, which often occurs in mixed infections with *Fusarium semitectum* and other fungi (Knight et al. 1977). *C. musae* also infects other injuries on bananas, particularly those on the ridges of angular fruit, causing decay which spreads into the pulp during ripening. Anthracnose (*C. musae*) gets its name from the ability of the fungi to cause latent infections of banana fruit. These infections occur at any time during the development of the fruit in the field, and usually only develop as the fruit ripens, causing round lesion on the peel surface that can cause rotting in the pulp during ripening. This was a problem when bananas were shipped as bunches with prolonged shipping times, or when ripened at temperatures above 18 °C. It is rarely seen in hands packed in boxes. For international trade bananas are treated postharvest with a chemical fungicide (usually Thiabendazole or Imazalil) to control Crown Rot and other fungal diseases (Fig. 2.3) although other non-fungicidal methods have been tested including covering the cut areas of the crown with cling film (Fig. 2.4).

Other fungi associated with postharvest diseases of banana fruit include:

Botryodiplodia theobromae Stalk and Fruit Rot,
Ceratocystis paradoxa Stem End Rot,
Magnapothe grisea, *Pyricularia grisea* Pitting Disease,
Verticillium dahliae Cigar End Rot.

Most of these diseases are only occasionally serious where infection levels are high and favourable conditions occur; none are as widespread as Crown Rot. Pitting Disease, however, is a serious field problem in some production areas. It can also be serious problem on fruit after harvest due to the development of latent infections that are not controlled by postharvest treatments (Meredith 1963; Stover 1972). All postharvest disease organisms are widespread in the field, growing and sporulating on decaying banana flowers, bracts and leaves. The spores are blown by wind or

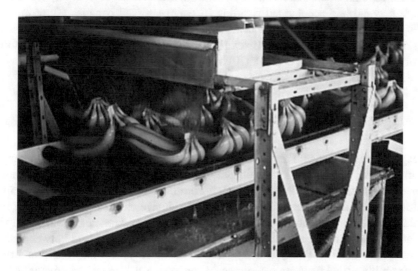

Fig. 2.3 Cascade application of a fungicide to freshly harvested 'Valery' in Ecuador, before being packed for export in 1996

Fig. 2.4 Plastic covering
for control of crown rot
disease on organic
'Cavendish' bananas,
applied directly after
harvest. Photograph
taken in 2013 after
importation in UK and
ripening

splashed by rain onto the fruit and can also be carried on bunches to contaminate the
packing station environment, including the washing water. Daundasekera et al.
(2008) showed that *C. musae* can produce ethylene *in vitro* on bananas. They found
that *C. musae* isolate CM100 was capable of producing ethylene *in vitro* on
methionine-supplemented basal medium and on banana peel extracts that contained
methionine.

Water Stress

Water stress can induce increased ethylene production (Kubo et al. 1990) in bananas,
which may explain accelerated ripening in water-stressed bananas (Burdon et al.
1994). Karikari et al. (1979) reported that 'Apem' responded to water stress by
marked reduction in their pre-climacteric period, which is consistent with previous
reports by Sanchez-Nieva et al. (1970) and Thompson et al. (1974).

Damage

In experiments in Ghana, Ferris et al. (1993, 1995) showed that mechanical damage
promoted early ripening in 3 plantain genotypes: 'Ubok Iba' (True Horn), 'Agbagba'
(False Horn) and 'Obino I'Ewai' (French Plantain) when harvested at either fully
mature or immature stages. The mechanical damage that caused early ripening
could be due to impact, abrasion or incision. The effect of damage on plantain ripening affected all three genotypes but the intensity of the effect varied between genotypes. Ferris et al. (1993a) and Bugaud et al. (2014) also showed that bruise
susceptibility varied between genotypes, but as long as the impact energy was below
200 mJ, neither 'Cavendish' or 'Flhorban 925' were susceptible to bruising.

Bunch Covers

Banana bunch covers are plastic film tubes that are used to cover bunches shortly after they have emerged from the top of the pseudostem (Fig. 2.5). This is to protect the fruit from dust, bird dropping, leaf rubbing and scarring but particularly from damage by insects particularly different types of thrips (*Thrips hawaiiensis, Chaetanaphothrips signipennis* and *Hercinothrips bicinctus*). Sometimes bunch covers are coated with an insecticide to improve control of thrips. Bunch covers can also have postharvest effects. After harvest the pre-climacteric life of the covered bunches was one or 2 days longer than the ones that had been grown with no covers. Johns and Scott (1989) explained this effect as the fruit under the covers were less exposed to the environment, therefore they lost less water and had a higher moisture content, which gave the appearance of being more mature, since they would be more rounded (see section on Harvest maturity). Choudhury et al. (1996) also reported a longer green life for covered 'Dwarf Cavendish' fruits whereas Parmar and Chundawat (1984) reported a reduction in green life using blue polyethylene covers on 'Basarai' bananas. After harvest they found that the fruits grown under a cover lost more weight after being harvested than those that had not been covered. They attributed this to be probably as a consequence of going from a higher humidity environment during growing to a lower humidity environment after harvest. Parmar and Chundawat (1984), on the contrary, recorded lower postharvest weight losses in fruits that had been grown under bunch covers. They also found, with regard to quality, that the fruit that had been covered, except for those under non-transparent covers, had a higher TSS content than those not covered. They

Fig. 2.5 The protective plastic sleeve is tied in a knot at the bottom but a gap is left to allow water to escape and a coloured ribbon (in this case red) is tied on to identify the age of the bunch

conjectured that this was probably because the higher temperature under the cover favoured the conversion of starch into sugars. The reduction in the content of TSS in fruits grown under non-transparent covers might also be due to the higher moisture content of these fruits. The sugar to acid ratio of the covered fruits was also higher than those not covered fruit.

Harvest Maturity

The maturity at which bananas are harvested clearly affects their quality and ripening (see Chap. 3 section: Maturity of fruit and response to ethylene). Various ways are used by growers to assess harvest maturity, but the most commonly used in international trade is "ribbon tagging". The time between flowering and fruit being ready for harvesting may be quite constant, so in order to identify exactly when anthesis occurred a coloured plastic ribbon is attached to the bunch directly after anthesis, usually together with a plastic bunch cover (Fig. 2.5). The same colour is used for 1 week and changed to another colour the following week and so on. This means that at the harvest time the age of is bunch is precisely known. Wilson Wijeratnum et al. (1993) showed that the 'Ambul' bananas grown in Sri Lanka reached physiological maturity 8–9 weeks after the flowers had opened. Fruit growth and development continued until the thirteenth week but changes in other physical and chemical parameters were minimal after 11 weeks. For 'Cavendsh' in Ecuador the maximum time from anthesis to harvest is usually 12 weeks and in the Windward Islands it is 13 weeks. Harvest maturity has been shown to affect eating quality of other fruit. For example, Blissett et al. (2019) compared 'East Indian' mangoes that had been allowed to ripen before and after harvesting. That is fully ripe golden yellow and yellow-orange skin harvested 120 days after anthesis and mature-green of a comparable size 112 days after anthesis and both harvest maturities were ripened at 27 °C. At eating ripeness, they found no significant difference in TSS but tree ripened mangoes had lower total reducing sugar, titratable acidity, higher total phenolics and higher DPPH (1,1-diphenyl-2-picrylhydrazyl) free radical scavenging activity than postharvest ripened of the mangoes.

The way that ribbon tagging works in practice is that the grower has options when he needs to harvest a bunch and chooses which bunches to harvest, not only on their age, but also their angularity and calliper grade. In a hypothetical example (Fig. 2.6) the grower can harvest only bunches with white, green, yellow of blue ribbons. Bunches with a white ribbon were 10 weeks old, bunches with a green ribbon 11 weeks old, bunches with a yellow ribbon 12 weeks old and bunches with a blue ribbon 13 weeks old. The grower must harvest all bunches with blue ribbons (if their calliper grade is still below 37 mm or over 46 mm they must not be packed for export), but the grower has the option of harvesting yellow, green and white depending on their angularity and calliper grade.

The thickness of individual fingers increases as the fruit matures and this thickness can be used to determine harvest maturity. It is called calliper grade and

Week beginning	Number of bunches harvested			
7 January	white	green	yellow	blue
14 January	brown	white	green	yellow
21 January	red	brown	white	green
28 January	purple	red	brown	white

Fig. 2.6 Simple chart to keep record of the number of bunches harvested and their age or maturity

Fig. 2.7 Plastic template for measuring maximum and minimum calliper grade for harvesting bananas for export in the Windward Islands in the 1970s

measures the width on an individual finger on a bunch. Callipers are used to measure the maximum acceptable width for fruit and the minimum acceptable widths for harvesting and various devices are supplied to growers (Fig. 2.7) to assist harvesters.

Because there is variation in maturity between hands on the same bunch, so in order to standardize the selection, calliper grade is measured only on a middle finger on the second hand from the top (Fig. 2.8).

Recommendations vary but commonly fruit with a calliper grade (diameter) of over 46 mm should not be packed for export and fruit less than 37 mm should not be harvested before 13 weeks after anthesis. Any bunch that still has not achieved a calliper grade of 37 mm after 13 weeks should be harvested and discarded or marketed locally, as the fruit may begin to ripen during transport.

The shape of each finger in cross section changes as the fruit matures. This change in shape is called angularity and is used, often in combination with ribbon tagging and other methods, to determine when to harvest a bunch. Various terms are used in producer countries to classify the angularity of banana fingers (Fig. 2.9). In some countries the marketing agents for export companies require farmers to harvest only a particular grade. In the Caribbean Islands grades were classified, for many years, from 'light three-quarters' to 'round full', which have long been used

Fig. 2.8 Diagram showing
the second hand form the
top of the bunch that
should be selected for
measuring calliper grade

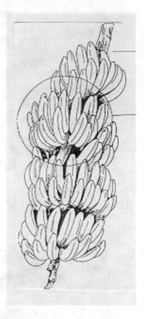

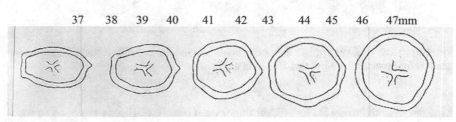

Fig. 2.9 Cross section of 'Cavendish' banana fingers showing the changes in shape of fruit as they grow and relating the shape to their width in mm

as the basis for judging when to harvest bunches for export (von Loesecke 1949; Stover and Simmonds 1987). These are based on a subjective combination of angularity and thickness of individual finger.

The maturity of hands in the bunch varies slightly; those at the proximal end of the bunch being more mature than those at the distal end, so the estimate of maturity is based on the fullness of fruit of the middle hand. In Latin America, where large-scale production is centred on plantation units of around 400 ha, in which uniform production conditions are achieved, maturity is assessed by the harvester using a specially designed spring calliper to measure the thickness of a standard finger in the bunch. This is usually taken as the middle finger of the outer whorl of the second hand from the distal end of the bunch. In some countries, farmers harvest less mature bananas, even for local markets, because they have cash-flow problems and they harvest as soon as the ripening room operator will accept the fruit (Thompson 1981).

Fig. 2.10 Showing a
banana finger that has split
postharvest, perhaps
because of a combination
of over maturity at
harvest and irrigation or
rainfall just prior to
harvesting

Ahmad et al. (2001) found very strong evidence for 'Robusta' that the fruit had much better organoleptic properties the more mature they were when harvested. However, if bananas are allowed to mature fully before harvest and harvesting is shortly after rainfall or irrigation, the fruit can easily split during handling operations (Fig. 2.10), allowing microorganism infection and postharvest rotting (Thompson and Burden 1995).

Chapter 3
Fruit Ripening

Generally, all fruits have been classified into either non-climacteric or climacteric on the basis of their respiratory pattern and ethylene evolution during ripening (Biale 1964). The ripening process is recognised as a complex irreversible scries of processes, which is initiated by a range of phytohormones primarily ethylene but also others such as abscisic acid (ABA) (Iqbal et al. 2017). It was previously thought that only climacteric fruit produced ethylene, but with the advent of precise analytical techniques it has been shown that all fruit and vegetables that were tested can both produce ethylene and react to exogenous ethylene. For example, Solomos and Biale (1975) showed that exposure to exogenous ethylene stimulated the respiration rate of many different types of fruit and vegetables (Table 3.1).

The rapid increase in endogenous synthesis of ethylene in climacteric fruit occurs at completion of fruit development and involves many physiological changes as well as increased respiration rate. In non climacteric fruit, the upsurge of respiration rate and ethylene biosynthesis were not observed or are transitory even after exogenous ethylene application (Tucker 1993; Duan, et al. 2007; Kays and Paull 2004).

It was contended by Obando et al. (2007) that the classification of fruit into climacteric and non-climacteric categories is an over-simplification. The simple distinction between climacteric and non-climacteric fruits has been queried because many fruits such as guavas, melons, Japanese plums, Asian pears and peppers show climacteric as well as non-climacteric behaviours depending on the cultivar. Studies on genetic and inheritance patterns indicate that ethylene-dependent and ethylene-independent pathways coexist in both climacteric and non-climacteric fruit (Paul et al. 2012). Zhang et al. (2009) contended that many non-climacteric fruit show a climacteric pattern of ripening. Also, Barry and Giovannoni (2007) contended that there was evidence that climacteric and non-climacteric fruits share some similar pathways of ripening. For example, in plums there are non-climacteric, suppressed-climacteric and climacteric cultivars. Some cultivars of plums ripen very slowly since ethylene production is suppressed and are referred to as suppressed-climacteric

A. K. Thompson et al., *Banana Ripening*, SpringerBriefs in Food, Health, and Nutrition, https://doi.org/10.1007/978-3-030-27739-0_3

Table 3.1 The effects of exposure to exogenous ethylene on respiration rate (μL O$_2$ g^{-1} h^{-1}) during storage of various fruit and vegetables compared to those not exposed to exogenous ethylene. Modified from Solomos and Biale (1975)

	Ethylene addition	
Crop	0	+
Apple	6	16
Avocado	35	150
Beet	11	22
Carrot	12	20
Cherimoya	35	160
Grapefruit	11	30
Lemon	7	16
Potato	3	14
Rutabaga	9	18
Sweet potato	18	22

cultivars. They may remain under-ripe if harvested too early. Other cultivars e.g. 'Early Golden' ripen very rapidly after harvest. Genetic analysis showed identical DNA profiles for the suppressed-climacteric cultivars 'Santa Rosa', 'July Santa Rosa', 'Late Santa Rosa', 'Casselman' and 'Roysum' and the novel non-climacteric 'Sweet Miriam' (Minas et al. 2015).

ABA content of climacteric fruit is very low at immature stages, but increases and peaks just before the onset of ripening and autocatalytic ethylene biosynthesis. Lohani et al. (2004) suggested that ABA stimulates ripening in bananas independently from ethylene. Exogenous ABA application was shown to enhance fruit softening, with exception to PG action. Jiang et al. (2000) suggested that ABA may act as a coordinator during the climacteric period in bananas by enhancing their sensitivity to ethylene. Ethylene and ethylene mediated responses suggest that sensitivity towards ethylene differs in various tissues and in different developmental stages of fruit, because of signalling interactions of ethylene with other plant hormones, metabolites and environmental signals (Stepanova and Alonso 2005; Kendrick and Chang 2008; Yoo, et al. 2009; Zhang et al. 2009).

Pre-Climacteric Phase

Generally, bananas are harvested at the pre-climacteric stage (colour index 1, Table 1.1; Fig. 1.1), which is prior to the autocatalytic generation of ethylene. During the pre-climacteric phase, the peel of the fruit is still green, the principal component in the pulp is starch (approximately 85 % of pulp dry weight) (Cordenunsi and Lajolo 1995) and it has a relatively low basal respiratory rate of about 10–20 mg CO$_2$ kg h^{-1} and undetectable ethylene production (Chillet et al. 2008). The low respiration rate of pre-climacteric bananas is concomitant with very low fructose-1,6-bisphosphate content (Ball et al. 1991). Sucrose content in the pre-climacteric fruit

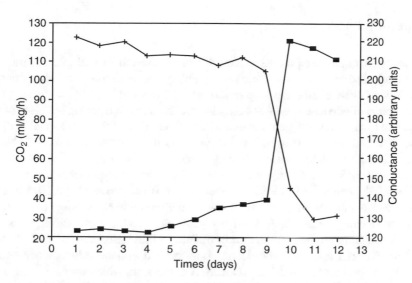

Fig. 3.1 Changes in CO_2 evolution (■) and in the conductance (+) of intact fruit during the pre-climacteric and climacteric of 'Giant Cavendish' bananas. (From Nolin 1985)

pulp increases slowly due to the very low activity of sucrose phosphate synthase (Cordenunsi and Lajolo 1995). During the pre-climacteric period, ethylene production of bananas was reported to be undetectable (Liu et al. 1999; Zhang et al. 2006; Duan, et al. 2007; Larotonda et al. 2008). Liu et al. (1999) reported that in unripe bananas, the expression of the *MA-ACS1* gene is very low, which is concomitant with the undetectable ACS activity whilst ACO activity and abundance of the *MA-ACO1* gene increase gradually during the pre-climacteric period. Thus, the very low levels of the both ACC content and ACS activity in pre-climacteric bananas limit the initiation of autocatalytic ethylene biosynthesis (Pathak, et al. 2003).

The duration of the pre-climacteric stage can be estimated by measuring the grade of the fruit (see Chap. 2, Harvest maturity), but greater precision can be obtained by measuring the minimum concentration of ethylene that can initiate the climacteric (Nolin 1985). The end of the pre-climacteric phase can be determined by measuring the increased softness of the pulp (New and Marriott 1974; Marriott and New 1975; Marriott et al. 1979). Texture can be determined subjectively by touch or objectively by the conductance of the peel and pulp and by correlating this conductance with the rate of respiration (Nolin 1985; Marchal et al. 1988) (Fig. 3.1). The dielectric constant and loss factor of material correlated well with moisture content and ripeness of fruit (Soltani et al. 2011). They found that the best frequency of sine wave, for predicting the level of ripeness in bananas was 100 kHz, which gave a coefficient of determination (R^2) of 0.94. Further factors, besides the physiological state, influence the duration of the pre-climacteric, including the temperature, humidity and atmospheric composition.

Ripening Phase

Ripening is a complex process of coordinated activation of multiple genetic and biochemical pathways. Climacteric fruit ripening involves a transient rapid increase in ethylene biosynthesis and respiration rate. Davies et al. (2006) suggested that "perception of ethylene is vital not only for the initiation of ripening but also for the continued expression of genes required for ripening". McMurchie et al. (1972) described two systems of ethylene biosynthesis that operate in bananas. During maturation there is a low basal rate of ethylene biosynthesis termed System 1. This is followed by System 2 ethylene biosynthesis, which is responsible for the auto-catalytic climacteric rise in ethylene production. Bouzayen et al. (2010) reported that *LeACS6* and *LeACS1A* are expressed at the pre-climacteric stage (System 1), while at ripening initiation, *LeACS4* and *LeACS1A* are the most active genes and *LeACS4* continues to be express during ripening while the expression of *LeACS1A* declines. The rise in ethylene production is associated with the induction of *LeACS2* and the inhibition of *LeACS6* and *LeACS1A* expression, which may be critical for the switch from System 1 to System 2. At the onset of ripening, bananas undergo a rapid burst in ethylene production, followed by a marked increase in respiratory rate (Liu et al. 1999). Afterwards, the irreversibly ripening-associated mechanisms such as the degradation of chlorophylls in fruit peel, the increase in conversion of starch to sucrose, the reduction of peel thickness, the increase in leakage in pulp electrolyte and volatile production occur (Wade 1995; Cordenunsri and Lajolo; 1995; Golding et al. 1999; Youryon and Supapvanich 2017). Unlike most climacteric fruit, ripening in bananas is characterized by a biphasic respiratory climacteric which proceeded by ethylene evolution (Pathak et al. 2003).

Internal Ethylene

Unlike most other climacteric fruits, a huge production of ethylene in bananas starts at the onset of climacteric phase and then rapid falls during the rise of the respiration rate (Karikari et al. 1979; Duan et al. 2007). Peak ethylene production in some climacteric fruit was given by Belitz et al. (2009) as: avocados 500 µg L^{-1}, bananas 40 µg L^{-1}, pears 40 µg L^{-1}, tomatoes 35 µg L^{-1} and mangoes 3 µg L^{-1}. The change in physiology of climacteric fruit from maturation to ripening is initiated when cellular quantities of ethylene reach a threshold level (Yang and Hoffman 1984). The metabolic pathway within the cells of higher plant that result in ethylene production involves a series of steps culminating in the conversion of methionine to SAM (S-adenosyl-methionine) which is converted to ACC (1-aminocyclopropane-1-carboxylic acid) by the action of ACC synthase. ACC is the immediate precursor of ethylene biosynthesis in plants. ACC oxidase is required to convert ACC to ethylene (Hamilton et al. 1990). In the pre-climacteric phase of banana development, ethylene production is very low and starts to increase at the onset of climacteric phase (Duan et al. 2007). Previous work in bananas has shown that the induction of ethylene biosynthesis is primarily regulated by the expression of *ACS* and *ACO* genes (Liu, et al. 1999; Zhang et al. 2006). They found that among *MA-ACS* genes, *MA-ACS1* was the key ripening gene in bananas playing a crucial role in ethylene biosynthesis regulation during ripening, which is the only gene expressed during ripening. Whereas *ACO* activity increased gradually during the pre-climacteric phase and then there was an abrupt increase in *ACO* activity in parallel with the burst of ethylene biosynthesis (Liu et al. 1999). Karmawan et al. (2009) identified nine *ACS* gene family members in *Musa* AAA, called *MA-ACS1–9*, but only *MA-ACS1* correlated with fruit ripening. Lòpez-Gòmez et al. (1997) found the large increase in ethylene biosynthesis in bananas during ripening was also associated with the increased expression of *MA-ACO1* gene and *ACO* activity. After the climacteric peak, ethylene production decreased during the increase in respiration rate. The decrease in ethylene biosynthesis, associated with a rapid reduction of *ACO* activity, was limited through the decline of its cofactors (ascorbate and iron) or other unknown factors (Liu et al. 1999). Xuejun Liu et al. (1999) ripened 'Grand Nain' bananas at 22 °C naturally or after treatment with 100 µL L^{-1} ethylene for 18 hours. They also found that ethylene synthesis was regulated by transcription of *MA-ACS1*, but only until the climacteric rise in respiration and may subsequently be limited by accessibility of cofactors *in situ* for example ascorbate and iron. They showed that the typical pattern of ethylene biosynthesis when bananas are allowed to ripen without the application of exogenous ethylene (Fig. 3.2) was changed when ripening was initiated with exogenous ethylene. (Fig. 3.3). Cyanide is produced along with ethylene, but does not accumulate in the oxidative breakdown of ACC that were catalysed by the ethylene forming enzymes (Blackbourn et al. 1990;

Fig. 3.2 Changes in ethylene biosynthesis and 1-aminocyclopropane 1-carboxylic acid production in bananas ripened naturally. Modified from Xuejun Liu et al. (1999)

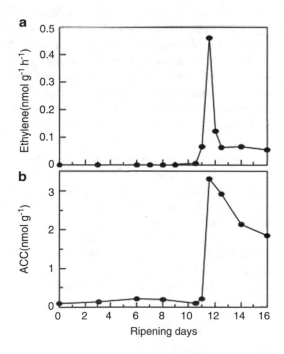

Fig. 3.3 Changes in
ethylene biosynthesis and
1-aminocyclopropane
1-carboxylic acid
production in bananas
initiated to ripen with
exogenous ethylene.
Modified from Xuejun Liu
et al. (1999)

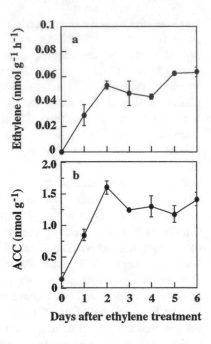

1990a). Ke and Tsai (1988) found that during ripening the ACC concentration in the pulp was higher and the ethylene-forming enzyme activity lower compared with the peel. Ethylene production was mainly from the pulp. In the absence of the pulp, the peel did not degreen fully unless supplied with exogenous ethylene. They concluded that degreening of the peel relied on ethylene diffusing from the pulp to the peel.

Pathak et al. (2003) showed that bananas have a "unique biphasic respiratory climacteric". Working with 'Dwarf Cavendish' they found that ethylene evolution on day 1, after the application of exogenous ethylene to pre-climacteric bananas, there was a 8.3 times increment in ethylene evolution followed on day 2 by a ten fold increase in respiration rate and on day 4 there was a 6.9 fold increase in ethylene evolution followed by a six fold increase in respiration rate on day 6. They suggested that this second peak justifies the appearance of a new ethylene binding site and the synthesis of an ethylene receptor in banana pulp. The second ethylene production in bananas is concomitant with the breakdown of starch resulting in the accumulation of sugars and flesh softening in the post-climacteric peak stage (Yang and Ho 1958). The starch breakdown during banana ripening is associated with the increase in sucrose phosphate synthase activity and the decline of sucrose synthase activity (Cordenunsi and Lajolo 1995). The sucrose phosphate synthase activity and sucrose synthesis in bananas is strongly stimulated by ethylene (Hubbard et al. 1990). During the post-ripening initiation phase, water content and soluble pectin of banana pulp increased gradually (Yang and Ho 1958), which is consistent with softening being associated with cell wall hydrolases. Ethylene evolution during banana ripening plays a key role in up-regulating PME, PG, pectate lyase and cellulase (Lohani et al. 2004).

Effects of Ethylene Post Ripening Initiation

Exposure to exogenous ethylene has been shown to hasten the ripening of mangoes, even after they had been initiated to ripen (Burg and Burg 1962), but no reports of similar effects could be found for bananas. One supermarket chain in Thailand used ethylene absorbing sachets in their MAP bananas in an attempt to increase their shelf-life, both whilst being offered for sale on their shelves and for customers at home. The temperature is relatively high in both places and even under air-conditioning the temperature can be about 25 °C. It was reported that this treatment was unsuccessful in extending the shelf-life of the bananas and it was not used in commercial practice by that company (Patcharin Chitaurjaisuk personnel communication 2019). It was suggested that this was not surprising since these bananas would have been initiated to ripening by ethylene treatment before they were packed and therefore absorbing any ethylene in the pack that was produced by the bananas would have no effect on speed of ripening. This hypothesis is supported by Golding et al. (1998), working with 'Williams', who found that their respiration rates, peel colour and total volatiles production, "once engaged with autocatalytic ethylene production, become partially independent of further ethylene action".

However, slowing the ripening of bananas that had been initiated to ripen can be achieved by 1-MCP treatment or changing the atmosphere around the fruit. Bananas treated with 1-MCP, after they had been initiated to ripen, had an extended shelf-life of 4–6 days compared to those not treated (Anonymous 2019) (See also Chap. 4 1-Methylcyclopropene). Liu (1976a) showed that bananas that had been initiated to ripen by exposure to exogenous ethylene and then immediately stored in 1 kPa O_2 at 14 °C remained firm and green for 28 days but then ripened almost immediately when transferred to air at 21 °C. Ahmad and Thompson (2006) showed that the marketable life of 'Giant Cavendish' could be extended 2.3–3.8 times, depending on the combination of $O_2 + CO_2$ used, compared to storage in air. However, they found that there were detrimental effects on fruit quality when 2 kPa O_2 was used and overall 4 kPa O_2 was most effective in extending their storage life. CO_2, in the range tested (4–8 kPa CO_2) appeared to have no positive or negative effects on marketable shelf-life or fruit quality. Klieber et al. (2002) also found that exposure of 'Williams' to O_2 below 1 kPa at 22 °C after ripening initiation induced serious skin injury that increased in severity with increasing exposure over the times tested (6–24 h) and also did not extend their shelf-life. 'Sucrier' that had been initiated to ripen with ethylene and then placed in polyethylene film bags (0.03 mm) at 20 °C showed inhibited ripening and had a "fermentation flavour" (Romphophak et al. 2004). Storage of 'Williams', which had been initiated to ripen, in total N_2 at 22 °C had a similar shelf-life to those stored in air, but areas of brown discolouration appeared on the peel of bananas in total N_2 storage (Klieber et al. 2002). Hypobaric storage can slow ripening of bananas after they have been initiated to ripen. Liu (1976c) initiated 'Dwarf Cavendish' to ripen by exposure to 10 μL L^{-1} ethylene at 21 °C. They then stored them at 14 °C for 28 days under hypobaric storage of 51 or 79 mm Hg. All fruit remained green and firm for the 28 days and continued to ripen normally after they had been removed to atmospheric pressure at 21 °C (760 mm Hg). See also Chap. 4 Hypobaric storage.

Controlled atmosphere storage of bananas before ripening initiation has been shown to affect ripening. Wills et al. (1998) showed that pre-climacteric bananas exposed to low O_2 took longer to ripen when subsequently exposed to air than fruits kept in air for the whole period. 'Cavendish' that had been stored for 7 days in 0.5 or 2 kPa O_2 then transferring to air and initiated to ripen with ethylene, showed a delay in the onset of the climacteric rise in respiration rate and delays in yellowing of peel, softening and conversion of starch to sugars compared to the control that had been stored in air throughout (Imahori et al. 2013).

Genetic Effects on Ripening

The elements of the postharvest environment that have been shown to interact with genetic factors in fruit include temperature, ethylene, O_2 and CO_2. Ripening, textural and colour changes and susceptibility to diseases have been shown to be influenced by gene expression (Giovannoni 2001). Banana genes associated with both peel and pulp ripening changes have were described by Elitzur et al. (2010) and Manrique-Trujillo et al. (2007) who showed that differential gene expression during ripening in bananas was linked with changes in both metabolism and components. Seymour et al. (2011) reported that in strawberries, a MADS-box *SEPALLA* total acidity gene (*SEP1/2*) was needed for normal fruit development and ripening and in bananas the MADS-box *SEP3* gene also displayed ripening related expression (Elitzur et al. 2010). They characterized two banana E class (*SEPALLATA3*) MADS-box genes, *MaMADS1* and *MaMADS2*, which were homologous to the tomato RIN-MADS ripening gene (Seymour et al. 2013). The delay in fruit ripening in bananas was shown to be associated with reduced biosynthesis of ethylene, and in the most severe repressed lines, no ethylene was produced and ripening was most delayed in those lines. However, unlike tomato *rin* mutants, bananas of all transgenic repression lines responded to exogenous ethylene and ripening normally (Elitzur et al. 2016).

A number of genes involved in ethylene signalling have been identified mainly from the isolation of ethylene response mutants in *Arabidopsis*, defining a pathway from ethylene perception to changes in gene expression. Johnson and Ecker (1998) found a negative regulator of ethylene responses, *CTR1,* acting downstream of the ethylene receptors. This pathway suggests that the ethylene signal is propagated through a mitogen-activated protein kinase cascade. Yan-chao Han et al. (2016) showed a significant induction of *MaC2H2–1/2*, C2H2 zinc finger proteins, transcripts during the ripening of bananas with three different ripening characteristics caused by natural, ethylene-induced and 1-MCP delayed treatments, which correlated well with ethylene production. Also, *MaC2H2–1/2* bound to the promoters of the key ethylene biosynthetic genes *MaACS1* and *MaACO1* and repressed their activities. They concluded that *MaC2H2–1/2* are transcriptional repressors and may mediate a finely tuned regulation of ethylene production during banana ripening, possibly via transcriptional repression of ethylene biosynthetic genes. Jian-fei Kuang et al. (2013) reported that EIN3 binding F-box protein (EBF) is an essential signalling component necessary for ethylene response. They isolated two EBF genes designated *MaEBF1* and *MaEBF2* from bananas and found that *MaEBF2* was enhanced by ethylene during fruit ripening, while *MaEBF1* changed only slightly. They concluded that *MaEBF* may be involved in banana ripening, at least partly via interaction with *MaEIL5*. In the nucleus, the EIN3 family of DNA-binding proteins regulates transcription in response to ethylene and an immediate target of EIN3 is a DNA-binding protein of the AP2/EREBP family (Alonso et al. 1999).

Both elevated CO_2 and reduced O_2 atmospheres have beneficial effects in controlling overall metabolic rate and ethylene production during the postharvest period of fresh fruit and vegetables (Wills et al. 2007). Song et al. (2015) suggested that high levels of CO_2 in stores can compete with ethylene for binding sites in fruits. Burg and Burg (1967) showed that in the presence of 10 kPa CO_2 the biological activity of 1 % ethylene in the surrounding atmosphere was abolished. Yang (1981) showed that CO_2 accumulation in the intercellular spaces of fruit acts as an ethylene antagonist. Romero et al. (2008) stored grapes under 20 kPa CO_2 + 20 kPa O_2 + 60 kPa N_2 or air for 3 days, then air for 15 days. The grapes that had been exposed to controlled atmosphere for 3 days activated the induction of transcription factors belonging to different families such as ethylene response factors, in particular *VviERF2-c*. Rothan et al. (1997) found that CO_2 induced the expression of certain stress-related genes and blocked the expression of both ethylene-dependent and ethylene-independent ripening associated genes. Song et al. (2015) stored bananas and plantains in high CO_2 concentration (20 kPa) with O_2 at 21 kPa for 6 days at 24 °C. They found that storage in high CO_2 atmospheres suppressed ethylene evolution and the expression of the related biosynthesis gene, *ACS*, as well as up-regulated the expression of anaerobic respiration pathway genes, *ADH* and *PDC*, which might be responsible for the retardation of the pulp ripening.

In climacteric fruits the final step in the biosynthesis of ethylene requires O_2 (Adams and Yang 1979), therefore, a lower O_2 concentration might result in an inhibition of ethylene biosynthesis and slow the ripening process. Also, all fresh fruits and vegetables synthesize O_2, many in very low levels, so the effects on ethylene biosynthesis may also be a factor. Loulakakis et al. (2006) concluded from their work on low O_2 storage of avocados that the "low oxygen, in addition to its inhibitory effect on ethylene biosynthesis and action, exerts its effect on prolonging the storage life of fruit by inducing a number of genes and proteins which possibly participate in the adaptation of fruit to low oxygen and by suppressing or by un-affecting others without the involvement of ethylene". Tonutti (2015) contended that members of the ethylene responsive factor VII (transcription factors gene family) displayed differential expression suggesting their involvement in the modulation or controlling mechanisms where there is O_2 deficiency in apple cells. Imahori et al. (2013) showed that bananas stored in an atmosphere containing 0.5 kPa O_2 there was increased alcohol dehydrogenase activity and after the bananas were transferred to regular (21 kPa O_2) air and initiated to ripen with ethylene, the onset of climacteric peak, peel yellowing and pulp softening were delayed. Kanellis et al. (1989a; 1998b) found that storing avocados, which had been initiated to ripen, for 6 days in an atmosphere containing 2.5 kPa O_2 suppressed cellulase activity immune-reactive protein and abundance of its mRNA and also produced an alteration in the profile of total proteins, which involved suppression, enhancement and induction of new polypeptides (Kanellis et al. 1989a and 1993). In addition, it has been shown that O_2 levels of 2.5–5.5 kPa, which suppressed the appearance of ripening enzymes at the protein and mRNA levels, was similar to those O_2 levels that induced the synthesis of new isoenzymes of alcohol dehydrogenase (Kanellis et al. 1993).

Pre-storage treatments can also affect the genetic response of fruit to environmental conditions. For example, Kangliang Sheng et al. (2018) found that several key genes involved in phenylpropanoid, flavonoid and stilbenoid pathways, including *PAL*, *CHS*, *F3H*, *LAR*, *ANS*, *STS*, showed increased expression in response to UV-C treatment in grapes. Storage temperature, prior to ripening, can influence the organoleptic characteristics of ripe bananas through genetic expression. Xiaoyang Zhu et al. (2018) stored bananas at 5 °C, 13 °C or 20 °C and found that both 13 °C and 5 °C affected volatile-related amino acid and biosynthetic precursors of fatty acid compositions, although at 13 °C the production of only a few specific volatiles were reduced compared to 20 °C. The expression levels of the biosynthesis-related genes associated with volatiles, *MaHPL*, *MaLOX*, and *MaAAT*, were repressed at both 13 °C and 5 °C particularly 5 °C, while *MaADH* and *MaPDC* were up-regulated. They concluded that storage at 5 °C reduced volatile production by regulating different key enzymes and genes involved in volatiles biosynthetic pathways and were mainly mediated via the repression of the lipoxygenase and amino acid metabolic pathways.

Maturity of Fruit and Response to Ethylene

The three main factors affecting the response of bananas to exogenous ethylene are harvest maturity, the period from harvest until ethylene exposure began and the duration of exposure to ethylene. Liu (1976a, b and c) found that immediately after harvest, bananas do not respond to exogenous ethylene, even when exposed at high levels They reported that this effect could be due to the presence of inhibitors, which could explain the absence of ripening of bananas before harvest. Liu (1976b) found that bananas that are too young do not have the enzymatic systems required to synthesize ethylene. Basically, very low concentrations of ethylene (10–50 μL L^{-1}) are enough to initiate ripening of bananas, but, in commercial practice they may be exposed to 1000 μL L^{-1} of ethylene for 24 and sometimes 48 hours to ensure uniform ripening (Thompson and Burden 1995). For the same maturity the climacteric rise in respiration was initiated by increasing the concentration of ethylene until a threshold value of about 0.8 μL L^{-1} was reached, beyond which the response remained unchanged (Chang and Hwang 1990a, b and c). These effects depend on genotype, temperature and on the conditions in which they were grown. Inaba and Nakamura

(1986, 1988) have shown for 'Cavendish' that the minimum time of treatment with ethylene is a function of the concentration of ethylene supplied, the temperature of the fruit and the concentration of ACC just before treatment. Jedermann et al. (2015) found that ripening of 'Cavendish' from 14 week old bunches (weeks after flower emergence) did not differ from fruits of 15 week old bunches. They also found no ripening differences of the younger fruits from the bottom of the bunch (hand 7 and 8) compared to the older fruit at the middle (hand 4 and 5) or the top of the bunch (hand 1 and 2). In contrast, Ahmad et al. (2001) showed differences in ripening for 'Cavendish' from hands from different parts of the bunch and bunches of different ages but, they estimated that the ripening differences of the bunch position could not be perceived in a commercial ripening situation. They found that the time taken for fingers to be fully ripe after harvest was significantly ($p = 0.05$) affected by their age at harvest and also its position on the bunch (Table 3.2). Overall it can be interpreted that the older the individual finger at harvest the shorter the time needed for it to ripen. This was especially clear for fruit harvested after 10 weeks, but there was little difference between 12 and 14 weeks after anthesis.

The colour of the fingers, as measured with a colourimeter, at colour stage 6 (Table 3.3) varied significantly ($p = 0.05$), with fingers having less green and more yellow with increased harvest maturity and a similar trend from the lower to the upper position of the finger on the bunch (Table 3.3). It can therefore be inferred that the older the fruit at harvest the better its colour when it is fully ripe. Also, it illustrates that human perception of colour can be considerably different to instrumental measurement. However, Mustaffa et al. (1998) for 'Montel' bananas found that there were differences in size, weight, volume, peel colour, texture, TSS, AA, pH, TA, starch and sugar contents between different hands and different fingers from the same bunch. Ahmad et al. (2001) found that the texture of 'Cavendish' was firmer for fruits ripened at 13 °C than at 16 °C especially for the younger fingers (Table 3.4). This was also reflected in TSS which was lower in the fingers ripened at 13 °C than at 16 °C. Taken with the data in Table 3.4 and Table 3.5 it can be concluded that the fruit ripened at 13 °C were less ripe that those ripened at 16 °C in spite of the subjective assessment that they looked about the same colour. Also, TSS, firmness and instrumental measurement of colour reflect the stage of ripeness and there were no differences between the two temperatures. Subsequently Ahmad et al. (2007), working with 'Cavendish' bananas found that they ripened in 9 days at 18 °C and in

Table 3.2 Effects of harvest maturity and position of the hand in the bunch on the number of days to ripen to colour stage 6 (Table 1.1) at 13 °C after initiation to ripen with ethylene of 'Robusta' and 'Grand Naine'. Modified from Ahmad et al. (2001)

	'Robusta'			'Grand Naine'		
	Position of hand on bunch					
Time after anthesis	Top	Middle	Lower	Top	Middle	Lower
14 weeks	18	19	21	24	24	30
12 weeks	19	20	25	24	24	32
10 weeks	28	23	34	31	32	35

Table 3.3 Effects of harvest maturity and position of the hand on the bunch colour at colour stage 6 (Table 1.1) at 13 °C after initiation to ripen with ethylene. Modified from Ahmad et al. (2001)

	Negative $a*$ = increased greenness			Positive $b*$ = increased yellowness		
	Position of hand on bunch					
Time after anthesis	Top	Middle	Lower	Top	Middle	Lower
14 weeks	−3.7	−4.5	−4.7	49	51	49
12 weeks	−4.3	−4.6	−5.0	48	49	48
10 weeks	−4.9	−4.9	−5.1	48	49	47

Table 3.4 Effects of harvest maturity and position of the hand on the bunch on the texture and TSS at colour stage 6 (Table 1.1) after initiation to ripen with ethylene and storage at either 16 or 13 °C of 'Grand Naine'. Modified from Ahmad et al. (2001)

	16 °C			13 °C		
	Position of hand on bunch					
	Texture					
Time after anthesis	Top	Middle	Lower	Top	Middle	Lower
14 weeks	3.5	3.5	3.3	5.1	5.2	5.6
12 weeks	3.4	3.4	3.2	5.6	5.6	5.8
10 weeks	3.3	3.2	3.0	5.7	5.7	8.2
	TSS					
Time after anthesis	Top	Middle	Lower	Top	Middle	Lower
14 weeks	21.4	21.0	19.1	19.1	18.7	16.1
12 weeks	19.7	19.2	18.2	18.5	17.6	16.7
10 weeks	18.8	18.2	16.8	16.7	16.6	14.5

Table 3.5 Effects of harvest maturity and position of the hand on the bunch on taste panel scores (1 = low and 5 = high) at colour stage 6 (Table 1.1) after initiation to ripen with ethylene and storage of 'Grand Naine' at 16°C. Modified from Ahmad et al. (2001)

	Flavour			Sweetness			Acceptance		
	Position of hand on bunch								
Time after anthesis	Top	Middle	Lower	Top	Middle	Lower	Top	Middle	Lower
14 weeks	4.3	4.1	3.1	4.6	4.2	3.5	4.4	4.0	3.0
12 weeks	3.9	3.8	3.0	3.5	3.5	2.8	4.0	3.8	2.7
10 weeks	3.1	3.0	2.6	3.3	3.1	2.7	3.4	3.5	2.8

11 days at 16 °C, and there was a slight preference for flavour and acceptability for those ripened at 18 °C.

There were no significant differences ($p = 0.05$) in the taste panel assessments for ripe 'Grand Naine' ripened at either 13 °C (data not shown) or 16 °C (Table 3.5) although there were clear trends that the panellists preferred the more mature fruit in all the criteria used.

Factors Affecting Fruit and Response to Ethylene

Pathak et al. (2003) commented that "ethylene induced ripening of banana is characteristically different from that of other climacteric fruits and that ethylene biosynthesis may have more than one mechanism operating during ripening which are tightly controlled at various levels." Bananas have been shown to respond to ethylene concentrations as low as 0.015–0.05 μL L^{-1} at 21 °C (Liu 1976a), although concentrations greater than 0.001 μL L^{-1} were shown to be required for the initiation of the climacteric in avocadoes (Biale and Young 1971). The production of endogenous ethylene by different *Musa* genotypes placed in the same conditions varies. For example, Karikari et al. (1979) showed that under conditions where 'Cavendish' (*Musa* AAA) produced 2.1–2.5 μL^{-1} kg hr.$^{-1}$ ethylene, tetraploids (*Musa* AAAA) produced 6.7 μL^{-1} kg hr.$^{-1}$ ethylene and 'Apem' (*Musa* AAB) produced 34.8 μL^{-1} kg hr.$^{-1}$ ethylene. The internal factors regulating the response to ethylene may be associated with growth regulators (Brady 1987) and ethylene receptors could be involved, thus explaining the inhibition of ripening by silver, which is not overcome by exogenous ethylene as the silver is bound to the putative ethylene receptor (Saltveit et al. 1978). Ethylene-mediated regulation of starch-degrading enzymes at transcriptional and translational levels is crucial for sugar metabolism during banana ripening. Also, the crosstalk between ethylene and other growth substances, including IAA and abscisic acid, also influences primary sugar metabolism. The starch stored in banana pulp cells is compartmentalized within plastids, which also contain several common starch-degrading enzymes (Peroni-Okita et al. 2013; Junior et al. 2006; Purgatto et al. 2001).

Changes that Occur During Ripening

A general figure for the various chemicals in banana pulp were given by USDA (Table 3.6). Forster et al. (2003) showed that the chemical composition in banana pulp varied with the central part being higher in nutrients, except for ascorbic acid.

Peel Colour

The pigments in the peel of bananas and plantains are chlorophylls and carotenoids (Figs. 3.12; 3.13; 3.14). The change in colour of ripening fruits is associated with the breakdown of chlorophylls with carotenoid levels remaining relatively constant (Von Loesecke 1949; Seymour 1986; Montenegro 1988) (Fig. 2.2), however, this depends on the genotype. 'Cavendish' banana cultivars can fail to completely degreen when they are ripened at 24–25 °C and above (Seymour et al. 1987; Semple and Thompson 1988). This can result in bananas, which are ripe in every other

Table 3.6 The composition of banana fruit for 100 g^{-1} fresh weight (adapted from USDA 2012)

Component	Content	
Water	74.91	g
Energy	89	kcal
Protein	1.09	g
Total lipid (fat)	0.33	g
Carbohydrate, by difference	22.84	g
Fibre, total dietary	2.6	g
Total sugars	12.23	g
Calcium	5	mg
Iron	0.26	mg
Magnesium	27	mg
Phosphorus	22	mg
Potassium	358	mg
Sodium	1	mg
Zinc	0.15	mg
Total ascorbic acid	8.7	mg
Thiamin	0.031	mg
Riboflavin	0.073	mg
Niacin	0.665	mg
Vitamin B-6	0.367	mg
Folate, DFE	20	mcg_DFE
Vitamin B-12	0.00	μg
Vitamin A, RAE	3	mcg_RAE
Vitamin A, IU	64	IU
Vitamin E (alpha-tocopherol)	0.10	mg
Vitamin D (D2 + D3)	0.0	μg
Vitamin D	0	IU
Vitamin K (phylloquinone)	0.5	μg
Fatty acids, total saturated	0.112	g
Fatty acids, total monounsaturated	0.032	g
Fatty acids, total polyunsaturated	0.073	g

respect, remaining green. The higher the temperature the more obvious the effect. It is one cause of the physiological disorder of 'Cavendish' bananas called "pulpa crema" or "yellow pulp." This is where bananas are initiated to ripen on the plant, but because the ambient temperature is above 25 °C the pulp ripens but the chlorophyll in the skin is not fully broken down. Bowden et al. (1994) compared the peel colour of 'Cavendish' and 'Apple' bananas by measuring their $a*$ values in a Colour Difference Meter (the lower the $a*$ value the greener the peel). They found that their pre-climacteric readings were both just above −19, but after 28 days storage at 13 °C 'Cavendish' had gone to +2.2 and 'Apple' to −6.1. When they were subsequently initiated to ripen with ethylene at 19 °C, the $a*$ values were + 2.7 for

'Cavendish' and + 1.7 for 'Apple'. With plantains it was shown that complete chlorophyll destruction can occur even at 35 °C (Seymour 1985, Seymour et al. 1987). Studies on why this effect of temperature on degreening of 'Cavendish' bananas occurs failed to reach a definitive conclusion (Blackbourn et al. 1990), although there was some indication that it was related to the thylakoid ultrastructure of the chloroplasts (Seymour 1986, Blackbourn et al. 1990). Thylakoids are an internal system of interconnected membranes that are arranged into stacked (grana thylakoids) and non-stacked (stroma thylakoids) regions that are differentially enriched in photosystem I and II complexes. Blackbourn et al. 1990) showed that during ripening of 'Cavendish' at higher temperatures there was a retention of thylakoid membranes resulting in a relatively delayed breakdown in both chlorophyll b and chlorophyll a from the reduced dismantling of pigment-protein complexes. By contrast, degreening was complete within 4 days at both 20 °C and 35 °C in plantains (*Musa* AAB), where the thylakoid membranes and their associated pigment-protein complexes were lost, and there was rapid increase in chlorophyll a to b ratios at both ripening temperatures. They suggested that the retention of thylakoid membranes is an important factor in the failure of 'Cavendish' bananas to degreen when ripened at tropical temperatures, and that the degreening problem may be related to the comparatively high chlorophyll b content of the pre-climacteric fruit. Sultan and Rangaraju (2014) stored 'Grand Naine' at 20 °C and 95 % RH with 100 ppm ethylene for 24 h. Using the Hunterlab and CIE $L*a*b*$ systems they found that lightness ($L*$) and yellowness ($b*$) increased to 74.25 and 51.11 respectively by the 4th day and subsequently declined.

Colour changes during ripening of many fruits, including bananas, have been used as a rough guide to the stage of ripeness. It is commonly used commercially in the form of colour matching charts based on the degree of peel yellowing where colour index 1 (Table 1.1) is allocated to fruit that are dark green and colour index 6 (Table 1.1) is for fruit which are fully yellow (Von Loesecke 1949; Lizada et al. 1990; Thompson 1996). Examples of these colour charts are supplied by commercial banana companies and can also be found in Stover and Simmonds (1987), Abdullah and Pantastico (1990) and Thompson (1996).

Medlicott et al. (1992) measured changes in peel ground colour during ripening of bananas using peel colour scores, a colour difference meter and by extraction and measurement of chlorophyll and carotenoid concentrations. There were close correlations between the subjective colour scores and the colour meter measurements. A similar relationship was obtained for colour scores and chlorophyll content, but no significant correlations were found between colour scores and carotenoid content. Peel colour changes in relation to pulp firmness and TSS content were highly correlated and preceded simultaneously throughout ripening. Slaughter and Thompson (1997.) described an optical chlorophyll sensing system that showed a high correlation with spectral analysis, tristimulus colorimeter analysis and visual colour matching of the peel colour of bananas. This optical chlorophyll sensing system gave a rapid response, was simple to operate, non-destructive and low cost and was reported to have potential application in automatic monitoring of banana ripening.

Peel Spotting

Brown spots on the skins of bananas is a normal stage of ripening that usually occurs when the skin has turned from green to yellow or fully yellow depending on the genotype. Maneenuam et al. (2007) compared the effect of storage in different O_2 partial pressures at 25 °C and 90 % RH on peel spotting in 'Sucrier'. The fruit had first been initiated to ripen and were turning yellow. They were then transferred to atmospheres containing either 90 kPa O_2 or 18 kPa O_2 in gas tight chambers. Peel spotting and a decrease in dopamine levels were quicker in fruit in 90 kPa O_2 indicating that dopamine might be a substrate for the browning reaction. Dopamine levels in 'Cavendish' were between 80–560 mg 100 g^{-1} in the peel and 2.5–10 mg 100 g^{-1} in the pulp, even in ripened bananas (Kanazawa and Sakakibara 2000). The browning reaction is related to the activity of PAL that converts phenylalanine to free phenolic substances that form the substrate that is converted to quinones by PPO. The level of total free phenolics, both in the whole peel and in peel spots, was lower in the high O_2. They concluded that peel spotting was not correlated with *in vitro* PAL and PPO activities. Decreases in dopamine levels correlated with peel spotting, indicating that it might be used as a substrate for the browning reaction. Kamdee et al. (2018) ripened 'Sucrier' to colour index 3–4 (Table 1.1) and then placed them at 42 °C and 78 % RH for 6, 12 18 or 24 hours followed by storage at 25 °C and approximately 78 % RH. Exposure to 42 °C inhibited peel spotting but did not impair ripening, with exposure for 24 hours being the most effective, but the taste and odour of the ripe fruit was reduced to unacceptable levels. After the 18-hour treatment peel spotting was considerably reduced whilst the taste and odour remained acceptable. Total phenolics and dopamine content of treated bananas were higher and the *in vitro* activity of enzymes involved in the production of substrate (PAL) and in the initial browning reactions PPO was inhibited by exposure to 42 °C. The exposure to 42 °C apparently did not delay the membrane degradation required for the interaction between the substrate and the enzymes that catalyze the browning reactions, suggesting that the main effect of the heat treatment was in the enzymatic steps involved in the browning reactions.

Peel Chemicals

Changes in many chemicals that occur in banana peel during ripening have been dealt with above. Banana peel also contains substantial amounts of phytonutrients and minerals. Puraikalan (2018) found that powder made from banana peel contained total phenols (159–235 mg 100 g^{-1}), flavonoids (1.23–1.70 mg QE g^{-1}), tannin (127–152 mg 100 g^{-1}), protein (9.4–11.7 %), fat (3.6–6.7 %) and fibre(11.5–14.4 %), depending on variety, and could be incorporated into processed foods and serve as a functional food. Anhwange et al. (2009) analysed *M. sapientum* peel and gave the mineral content as potassium, calcium, sodium, iron, manganese, bromine, rubidium, strontium, zirconium and niobium to be 78.10, 19.20, 24.30,

0.61, 76.20, 0.04, 0.21, 0.03, 0.02 and 0.02 mg g^{-1}, respectively. The concentrations of protein, crude lipid, carbohydrate and crude fibre were: 0.9, 1.7, 59.0 and 31.7 %, respectively. They also commented that banana peel is a good source of nutrition and could be utilized as an animal feed supplement. Baskar et al. (2011) tested ethanolic extract of peel of free radical scavenging assays of peel extracts of nine local banana genotypes. They found that peel extracts of all the nine genotypes had significant antioxidant and phytochemical activity.

Finger Drop

Finger drop occurs during ripening and is associated with weakening of the pedicel, which can result in reducing the market value of the fruit. Sensitivity to finger drop was reported to vary between banana varieties (New and Marriott 1983). Esguerra et al. (2009) stored 'Latundan' bananas at 23–25 °C and found that finger drop occurred about 5–6 days after harvest when the fruit were fully yellow. A major cause of finger drop is infection, associated with crown rot fungi (see Chap. 2 Stress), but it has also been associated with physiological effects and was shown to be stimulated by ripening at higher temperatures (Semple and Thompson 1988). They also found that prolonged exposure of fruit to ethylene during ripening could increase finger drop in certain circumstances. Paull (1996) found that finger drop in 'Santa Catarina Prata' bananas occurred after the climacteric ethylene peak and was associated with a second ethylene production. Holding fruit at 15 °C for up to two weeks reduced finger drop of fruit ripened at 25 °C. Ripening fruit at 17.5 °C, or 1 day at 25 °C then at 17.5 °C reduced finger drop from up to 100 % to less than 10 %. Ethylene treatment for 1 day at 25 °C resulted in less finger drop than fruit that did not receive exogenous ethylene. More mature hands were more prone to finger drop than less mature hands. Finger drop was associated with enhanced ethylene biosynthesis gene expression including developmental related and ripening induced genes (*MaACO1*), specific ripening-induced genes (*MaACS1*) and wound-induced genes (*MaACS2*) (Mbéguié-A-Mbéguié 2008; Huber and Mbéguié-A-Mbéguié 2012). In studies of 'Cavendish' it was shown that a change in the expression of major cell wall modifying genes occurred in the finger drop area (Mbéguié-A-Mbéguié 2009). Spraying the bananas postharvest with 4% calcium chloride delayed the onset of finger drop by 2–4 days after the attainment of full yellow colour but treatment with GA$_3$ or ethanol did not control finger drop (Esguerra et al. 2009).

Weight Loss

Postharvest loss of moisture from the fruit can hasten ripening. It was found by Finger et al. (1995) that respiration rate and the biosynthesis of ethylene was higher when the fruits had fresh weight losses of 5 % or more. The increase in respiration rate was 70 % higher and ethylene production was 50 % higher for fruits that lost 5 % when compared to the control fruits in bananas which had lost little moisture. The results confirmed that water stress (after the harvest of fruit) might affect the shelf-life and ripening, depending on the intensity of the water stress.

Clearly the higher the humidity during ripening the lower will be the weight loss (Thompson et al. 1972). Thompson and Silvis (1975) found that 'Dwarf Cavendish' in the Sudan could lose as much as 27 % in weight in 0% RH compared to about 1% in 100 % RH in ambient temperatures (24-32 °C). Also packing fruit in plastic film or moist coir will reduce their weight loss. Generally, the more mature the bananas are at harvest the lower their weight loss during ripening with fully mature fruit loosing 5.2 %, less mature 6.3 % and for immature 10.4 % during 28 days at 16 °C (Ahmad et al. 2001). This trend was also true for plantains (Thompson et al. 1972) and bananas where weight loss varied in fruit from different positions on a single bunch (Table 3.5), but this could be confounded with speed of ripening since the fruit from the top of the bunch generally ripened quicker than those lower down. The temperature at which the fruit were ripened also affected their weight loss, with those ripened at the lower temperature (13 °C) losing more weight than those ripened at a higher temperature (16 °C) (Table 3.7). However, this might also be partly due to the fruit at the lower temperature taking longer to ripen than the ones at the higher temperature. Collin (1989) also showed considerable differences in weight loss during storage when comparing temperatures of 13.5 and 22 °C (Table 3.8).

Table 3.7 Effects of harvest maturity (time after anthesis) and position of the hand on the bunch on weight loss at colour stage 6 (Table 1.1) at 13 or 16 °C after initiation to ripen with ethylene. Modified from Ahmad et al. (2001)

	'Robusta' 13 °C			'Grand Naine' 16 °C			'Grand Naine' 13 °C		
	Position of hand on bunch								
Time after anthesis	Top	Middle	Lower	Top	Middle	Lower	Top	Middle	Lower
14 weeks	2.6	3.0	3.4	1.3	1.4	1.5	3.0	3.1	3.6
12 weeks	2.9	2.9	3.9	1.4	1.5	1.7	3.1	3.1	4.0
10 weeks	3.8	3.9	4.0	1.6	1.6	2.0	3.8	4.1	4.4
LSD ($p = 0.05$)		0.28			0.07			0.22	

Table 3.8 'Orishele' plantains harvested 82 days after flowering and stored for 21 days. Modified from Collin (1989)

Temperature	Weight loss %	Pulp:peel ratio	Acidity meq 100 g^{-1}	TSS %	Peel colour index
13.5 °C	8.9	2.2	8.1	22.1	6
22 °C	20.0	4.2	9.0	31.2	7+

Moisture

Cultural conditions and water availability during growth have a relatively limited effect on the water content of the fruit. The pulp of 'Cavendish' bananas had a higher water content than the pulp of plantains. Marchal and Mallessard (1979) showed that the moisture content of freshly harvested fruit was higher in the peel than the pulp ranging from 86.6 to 91.9 % for 'Cavendish' and 85.0–87.1 % for plantains in the peel and 60.3–74.2 % for 'Cavendish' and 60.4–60.6 % for plantains in the pulp. The increase in the weight ratio between pulp and peel reflects changes in moisture content. During ripening, the stomata, which were found to be equal in number in plantains and 'Cavendish' bananas (370–450 cm^{-2}), remain functional. A low atmospheric pressure initiates a degradation of the epidermal and stomatal tissues, thus reducing gaseous exchange including water vapour (Collin and Folliot 1990). During ripening water loss through transpiration from the peel, showed a pattern of changes similar to that of respiration (Palmer 1971). The rise in the water content of the pulp during ripening was linked to the breakdown of carbohydrates and osmotic migration of water from peel to pulp because of the higher concentration of sugars in the pulp (Fig. 3.4). This reflected the thickness of the pulp, which progressively became thinner during ripening (Fig. 3.5).

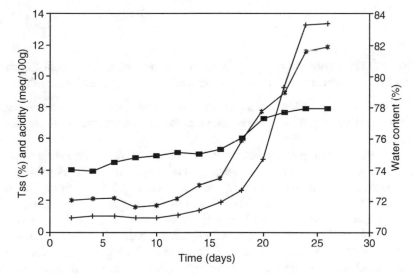

Fig. 3.4 Changes in the water content (∗), total free acidity (■), and in the total soluble solids (+) in the pulp of 'Giant Cavendish' bananas during the pre-climacteric and climacteric periods (Marchal et al. 1988)

Fig. 3.5 Changes in the
thickness of the peel (mm)
of three plantain genotypes
during ripening in Nigerian
ambient conditions of
21-33 °C and 79–99 %
RH. Modified from Ferris
et al. (1995)

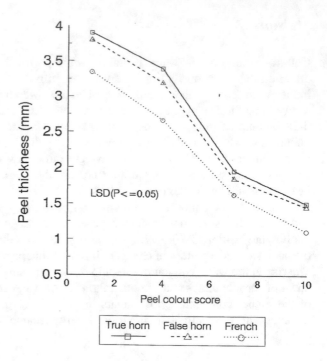

Texture

As bananas ripen both the peel and the pulp become softer. Softening of bananas
during ripening appears to be associated with two or three processes (Smith 1989):

1. The first is the breakdown of starch to sugars, which showed close correlation
   with softening (Fig. 3.6), since starch granules can have a structural function in
   cells.
2. The second is the breakdown of the cell walls due to the solublization of pectic
   substances and even the breakdown of cellulose. Smith found evidence of
   increased activity of cellulase during banana ripening. Changes in enzyme activity can be due to variations in the pH and of the fatty acid composition of the
   membranes. In tomatoes cellulase activity increased during softening (Huber
   1983), but the pulp pectinesterase activity declined during softening. However,
   there was no significant relationship, between the activity of pectinesterase or
   their starch content, or the soluble polyuronide levels (Smith and Seymour
   1990). Possibly these hydrolytic enzymes are induced by the increased concentration of endogenous ethylene, independently of the effect of ethylene on respiration (Brady 1987). Alterations in cell membranes increased their permeability
   and diminished the rupture force of both the peel and the pulp, with different
   genotype effects, in about 2 days before the climacteric rise. This could be linked

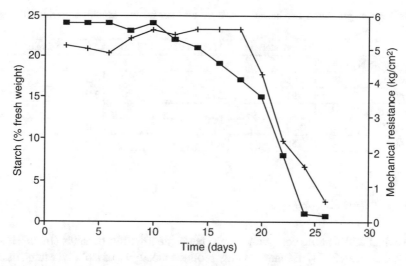

Fig. 3.6 Changes in the starch content of the pulp (■) and in the mechanical resistance of the fruit (+) during the pre-climacteric and climacteric in 'Giant Cavendish' bananas. (From Marchal et al. 1988.)

to a variation of water potential and of the membrane unsaturated fatty acids, which render the membranes more fluid and mobile (Wade et al. 1980). During softening the concentration of soluble pectic polysaccharides, uronic acid and related enzyme activities all increase, but these increases were not necessarily linked to the increase in soluble pectin.

3. The possible third process involved in softening is the movement of water from the peel of the banana to its pulp during ripening. This latter process could affect the turgidity of the skin which would be enhanced by transpirational losses. This change in the moisture status of the fruit also contributes to the ease of which the peel can be detached from the pulp.

Lohani et al. (2004) reported that softening during ripening is generally attributed to degradation in cell wall structure, particularly the solublization of pectin involving increased activities of various cell wall hydrolases. Their activity is regulated by ripening-related hormones and/or other signal molecules and exposure to endogenous ethylene stimulated the activity of PME, PG, pectate lyase and cellulase in 'Dwarf Cavendish' over a period of 7 days after ripening was initiated with ethylene. Pre-treating the fruit with either 1-MCP or IAA suppressed the ethylene effects with ABA stimulated activity of all hydrolases except PG and was most evident for pectate lyase. Therefore, ABA can enhance softening with or without ethylene. In contrast IAA suppressed their activity. Textural changes can vary with different banana genotypes with the peel of 'Cavendish' softening during storage and ripening while the peel of 'Apple' actually became slightly firmer (Table 3.9). For the pulp, softening was progressive but 'Cavendish' was consistently firmer than 'Apple'.

Table 3.9 Comparison of two banana genotypes on their texture. Modified from Bowden et al. (1994)

	Before storage	After 28 days at 13 °C	After ripening at 20 °C
Peel texture N mm²			
'Cavendish'	2.05	0.84	0.55
'Apple'	1.95	2.18	2.02
Pulp texture N mm²			
'Cavendish'	0.97	0.08	0.08
'Apple'	0.58	0.04	0.05

Flavour and Aroma

More than 250 volatile compounds have been identified in bananas (Imahori et al. 2013; Brat et al. 2004). Bananas mainly produce isoamyl and isobutyl esters, including 3-methyl butyl and 2-methyl propyl esters of acetate and butanoate. These esters are considered to be the major contributors to the aroma of bananas (Jayanty et al. 2002), providing the fruity smell (Wendakoon et al. 2006). The alcohols pentan-2-ol, 3-methyl butanol, and 2-methyl propanol also contribute to the succulent character of banana aroma and the ketone pentan-2-one, which is one of the major volatile compounds of bananas, imparts the characteristic banana flavour (Brat et al. 2004). McCarthy et al. (1963) and Tressl and Jennings (1972) had previously reported that amyl esters give bananas their distinctive flavour and aroma, and butyl esters give them a fruity flavour and aroma; however, other esters and aldehydes, alcohols, and ketones have been associated with flavour, and their production rates can increase during ripening. Salmon et al. (1996) showed that generally esters such as butyl acetate, isoamyl acetate, ethyl acetate, butyl butanoate and isoamyl isobutanoate are responsible for the characteristic aroma of fresh banana and constitute the major class of compounds present in banana's volatile profile.

Volatile compounds of several banana genotypes have been widely studied including 'Grand Naine' (Bugaud et al. 2009); 'Gran Enana', a subgroup of the 'Cavendish' originating from Central and South America (Vermeir et al. 2009); various varieties grown on Madeira Island (Nogueira et al. 2003); 'Cavendish' from Honduras (Jordan et al. 2001); free and glycosidically bound volatile compounds of 'Valery' and 'Pequeña Enana' (Pérez et al. 1997); 'Cavendish' 'Gran Enana' and 'Enana' from the Canary Islands, 'Enana' from Colombia (Cano et al. 1997); 'Cavendish', 'Sennin' and 'Delicious' (hybrid between 'Philippine' and 'Taiwanese') (Shiota 1993). Different volatilities and concentrations of those volatiles that can vary among the different varieties (Cano et al. 1997). They identified the compounds and the ratios of Spanish 'Enana', Spanish 'Gran Enana' and Latin American 'Enana' correlating with differences among the flavours with Spanish 'Enana' being the richest in flavour compounds. McCarthy et al. (1963) assigned the various components of banana aroma and flavour to the amyl and isoamyl esters of acetic, propionic and butyric acid, whereas the alcohols and carbonyls gave odours that they described as green, woody or musty. Only Spanish bananas exhibited the presence of hexyl butanoate, which was related to a banana-like flavour. Pontes et al. (2012) evaluated the volatile profiles of 'Dwarf Cavendish', 'Prata', 'Maçã', 'Ouro', and plantains and found a total of 68 volatile organic metabolites that were tentatively identified and used to profile the volatile composition in different *Musa* genotypes. Ethyl esters were found to comprise the largest chemical class, accounting for 80.9, 86.5, 51.2, 90.1 and 6.1 % of total peak area for 'Dwarf Cavendish', 'Prata', 'Ouro', 'Maçã', and plantain volatile fractions, respectively. Facundo et al. (2012) determined the volatile differences in two banana genotypes during cold storage and found that cold storage affected the volatiles more strongly in 'Nanicão' than in 'Prata'. Esters such as 2-pentanol acetate, 3-methyl-1-butanol acetate, 2-methylpropyl butanoate, 3-methyl butyl butanoate, 2-methylpropyl 3-methyl butanoate and butyl butanoate were considerably reduced during cold storage of 'Nanicão'.

In India Venkata Subbaiah et al. (2013) ripened bunches of 'Grand Naine' and found that their taste and flavour progressively increased up to the 6th day and thereafter decreased during the following 10 days. Tai-Ti Liu and Tsung-Shi Yang (2002) tested ripening of 'Pei-Chiao' bananas (a 'Giant Cavendish' clone) at 20, 25 or 30 °C over 8 days and found that development of flavour compounds was highest for those ripened at 25 °C and ethanol developed most rapidly at 30 °C.

Sanaeifar et al. (2014) tested a metal oxide semiconductor based electronic nose to monitoring the changes in volatile production of bananas during ripening and found that it was possible to distinguish between different ripening stages with 98.66 % accuracy.

Minerals

A general figure for the various minerals in bananas were given by USDA (Table 3.6) but, the mineral composition of all crops varies with cultural conditions and differences have even been observed between bananas and plantains harvested from the same plots (Table 3.10). Bananas have a high potassium content and Lee (2008) reported that an average sized banana had 450–467 mg K. The average K content for bananas grown in Hawai'i ('Dwarf Brazilian' and 'Williams') was 330.6 mg 100 g^{-1} fresh weight and Mg content averaged 35.1 mg 100 g^{-1}. Iron, copper, and manganese are other minerals of nutritional importance in bananas (Wall 2006). Hardisson et al. (2001) reported that for bananas grown in Tenerife there was 5.09 mg g^{-1} K, 0.59 mg g^{-1} P, 0.38 mg g^{-1} Ca and 0.38 mg g^{-1} Mg on a fresh weight

Table 3.10 Average mineral composition (mg 100 g^{-1} fresh weight) before ripening of pulp of 'Cavendish bananas' (*Musa* AAA) and plantains (*Musa* AAB) harvested from the same plot (modified from Marchal and Mallessard 1979)

	P	K	Ca	Mg	S
Bananas	27	460	7	36	34
Plantains	32	440	14	32	24

basis. Wall (2006) reported that 'Apple' bananas had higher concentrations of P, Ca, Mg, Mn and Zn than 'Williams'.

In Nigeria, Adeyemi and Oladiji (2009) gave the following for different stages of ripeness:

Unripe 73.47 % ash, 0.68 % Zn, 0.146 % Mn, 0.506 % Co.
Ripe 77.19 % ash, 0.80 % Zn, 0.271 % Mn, 0.886 % Co.
Overripe 79.22 % ash, 0.78 % Zn, 0.118 % Mn, 0.756 % Co.

Changes in mineral content during ripening must be related to the changes in mass of the fruit through water loss and to a lesser extent through mass losses due to respiration. The loss of water from the peel in the course of ripening can explain its enrichment in K, Ca and Mg (expressed in relation to their fresh weight). It is, however, possible that a proportion of these elements migrate with the water from the peel to the pulp, the K and Ca contents of which were shown to rise equally (Izonfuo and Omuaru 1988).

Carbohydrates

As bananas and plantains develop on the plant almost all the carbohydrate is in the form of starch, which accumulates during maturation, and in the pre-climacteric phase there is little change in the principal carbohydrate metabolites (Wardlaw et al. 1939). During subsequent ripening the total carbohydrate content is progressively reduced as the starch is broken down into reducing and non-reducing sugars (Barnall 1943). During the early part of ripening, sucrose is the predominant sugar, but in the later stages glucose and fructose predominate. Starch is broken down to sucrose by the action of sucrose phosphate synthetase and non-reducing sugars from sucrose by acid hydrolysis. In bananas the breakdown of starch is usually completed during ripening, but in plantains this breakdown is not complete even when they are con-sidered to be fully ripe (George 1981). In 'Apple' bananas (*Musa* AA) the fruit was found to still have residual starch when they were fully yellow and they may need to be ripened further before being eaten depending on the consumers' taste (Wei and Thompson 1993). To illustrate this point, the caption on the label of some *Musa* AA bananas, marketed in the UK, has "wait for brown freckles on skin before eating" (Fig. 1.3).

In bananas, ripening involves a reduction in starch content from around 15–25 % to less than 5 % in the ripe pulp, coupled with a rise of similar magnitude in total sugars (Barnell 1943, Desai and Deshpande 1975, Lizada et al. 1990). Wardlaw (1961) quoted figures that showed that the starch content of 'Gros Michel' could go as low as 1% and sugar content as high as 19% when they were fully ripe. Venkata Subbaiah et al. (2013) ripened 'Grand Naine' and showed that the total sugars increased from 1.22 % to 24.05 % and starch content decreased from 21.52 % to 2.53 %. Adão and Glória (2005) harvested 'Prata' bananas at the full three-quarter stage and stored them in polyethylene bags at 16 ± 1 °C and 85 % RH. Starch levels decreased throughout ripening, but after 7 days sucrose was prevalent and remained constant then decreasing after 28 days while fructose and glucose levels increased. However, at temperatures at or below 12 °C starch was no longer converted to sugar (Wardlaw et al. 1939). The onset of the starch-sugar conversion has been shown to be influenced by harvest maturity, with more mature fruits responding earlier. These changes have been demonstrated in both triploid (*Musa* AAA) (Madamba et al. 1977) and diploid (*Musa* AA) fruit (Montenegro 1988). Arora et al. (2008) reported that 'Karpuravalli' (*Musa* ABB) was rich in carbohydrates in terms of total starch (1786 μg g^{-1} dry weight in the peel and 545 μg g^{-1} dry weight in the pulp) and sugars (53.5 μg g^{-1} dry weight in the peel and 39.1 μg g^{-1} dry weight in the pulp).

Protein

Toledo et al. (2012) found that chitinases were the most abundant types of proteins in unripe bananas. Two isoforms in the ripe fruit were implicated in their stress/defence response and three heat shock proteins and isoflavone reductase were also abundant at the climacteric stage. Pectate lyase, malate dehydrogenase and starch phosphorylase accumulated during ripening. In addition to the ethylene formation enzyme amino cyclo carboxylic acid oxidase, the accumulation of S-adenosyl-l-homocysteine hydrolase was needed because of the increased ethylene synthesis and DNA methylation that occurred in ripening bananas.

Wade et al. (1972) reported that total nitrogen extracted by 5 % (w/v) trichloroacetic acid did not change significantly during ripening change and they found no evidence of a movement of nitrogen from the peel to the pulp during ripening. However, Dominguez-Puigjaner et al. (1992) working with 'Dwarf Cavendish' bananas found evidence of differential protein accumulation in the pulp during ripening.

Phenolics

Bananas can contain high levels of phenolic compounds especially in the peel (von Loesecke 1949). Tannins are perhaps the most important phenolic from the point of view of fruit utilisation because they can give fruit an astringent taste. As fruit ripens their astringency becomes lower, which is appearently associated with a change in the structure of the tannins, rather than a reduction in their levels, in that they form polymers (von Loesecke 1949, Palmer 1971). Mura and Tanimura (2003) confirmed the loss of astringency during ripening of bananas and also found that it was by the polymerizing of polyphenol compounds with a molecular weight of 2×10^5. However, Fatemeh et al. (2012) showed that green bananas had higher total phenolic compounds and total flavonoid compounds than those of ripe fruit. Radical scavenging activities (inhibition of DPPH) of the extracts ranged from 26.55 to 52.66 %. Phenolics are also responsible for the oxidative browning reaction when the pulp of fruit (especially immature fruit) is cut. The enzyme polyphenoloxidase is responsible for this reaction (Palmer 1971). Bennett et al. (2010) detected catechin, gallocatechin, epicatechin and condensed tannins, in the soluble extract of banana pulp but no soluble anthocyanidins nor anthocyanins. Fatemeh et al. (2012) evaluated the stage of ripeness (green and ripe) of both pulp and peel on antioxidant compounds and antioxidant activity of 'Cavendsh' and 'Berangan'. The TPC ranged from 75.01 to 685.57 mg GAE 100 g^{-1} and TFC ranged from 39.01 to 389.33 mg CEQ 100 g^{-1} of dry matter. TPC and TFC values of the peel were higher than those of the pulp. Although in 'Berangan' bananas the peel extracts appeared to have low TPC and TFC, its antioxidant activity was moderate to high, which implies that antioxidative compounds other than phenolics and flavonoids could be responsible for inhibition of DPPH.

Acidity

Bananas and plantains, like most other fruits, are acid with a pulp pH below 4.5 (Von Loesecke 1949). Palmer (1971) showed that the main acids in bananas were ascorbic, citric, malic and oxalic and the levels of these acids normally increase during ripening. Total acidity in the pulp was shown to increase rapidly from about 28 to 67 mL NaOH 100 g^{-1} fresh weight from the pre-climacteric minimum to the climacteric and then it slowly declined post-climacteric to about 52 mL NaOH 100 g^{-1} fresh weight (Wardlaw et al. 1939). These measurements were carried out on 'Gros Michel' at 29.4 °C and 85% RH over a ripening period of 14 days. Marchal et al. (1988) showed a slight increase in total acidity in 'Cavendish' during ripening (Fig. 3.4). Also, in 'Cavendish', the pH of immature fruit was reported to be between 5.4 and 6.0 both in the peel and in the pulp, but the free acidity of the pulp was higher (Thomas et al. 1986). During ripening pH decreased until it reached 4.0 at the fully ripe stage in both 'Cavendish' bananas and in plantains to increase gradually thereafter. In green fruits free acidity was lower in 'Orishele' plantain (AAB) at 2 meq 100 g^{-1} fresh weight (Collin and Dalnic 1991) than in 'Giant Cavendish' bananas at 4 meq 100 g^{-1} fresh weight (Marchal et al. 1988). Free acidity increased until full ripeness, more so in 'Orishele' (up to 10 meq 100 g^{-1} fresh weight) than in the 'Cavendish' (up to 7 meq 100 g^{-1} fresh weight). Collin (1989) reported that 'Orishele' plantains that had been stored for 21 days in air at 13.5 °C had acidity of 8.1 meq 100 g^{-1} but the acidity of those that had been stored in PE film bags was 4.5 meq 100 g^{-1}. A similar effect was shown at 22 °C where the figures were 9.0 and 2.9 respectively. Ruiling Liu et al. (2016) reported that 1-MCP maintained acidity in apples by regulating the balance between malate biosynthesis and degradation and also delayed the postharvest loss of malate and citrate. In mangoes Medlicott et al. (1987), Campbell and Malo (1969) and Fuchs et al. (1975) all indicated that exposing fruit to endogenous ethylene had little effect in increasing the rate of acidity loss, although Bhova et al. (1978) showed the converse, which may be a varietal effect.

Ascorbic Acid

Bananas are a good source of vitamin C with a general level of about 8.7 mg 100 g^{-1} (Table 3.6). Seenappa et al. (1986) reported that the vitamin C level was different in different varieties of banana and in 'Dwarf Cavendish' vitamin C level fell from 7.7 mg 100 g^{-1} in unripe fruit to 4.4 mg 100 g^{-1} in ripe fruit (Table 3.11) while the vitamin C levels for 'Cavendish' ranged from 2.1 to 18.7 mg 100 g^{-1} (Leong and Shui 2002, USDA 2012, Vanderslice et al. 1990, Wills et al. 1984). Also, in pre-climacteric fruit there was considerable variation in vitamin C content in different genotypes (Table 3.12). These results agree with Wenkam (1990), who reported vitamin C values of 5.1 mg 100 g^{-1} for 'Williams' and 14.6 mg 100 g^{-1} for 'Dwarf Brazilian'. Wills (1990) reported that 'Sugar' bananas (*Musa* AAB) had higher vitamin C, starch, glucose, fructose and dietary fibre than 'Cavendish'. Wall (2006) reported the following analysis: vitamin C 12.7 mg 100 g^{-1}, vitamin A 12.4 mg RAE 100 g^{-1}, total soluble solids 17.9 % and moisture 68.5% for 'Dwarf Brazilian'. For 'Williams' she reported: vitamin C 4.5 mg 100 g^{-1}, vitamin A 8.2 mg RAE 100 g^{-1}, soluble solids 20.5% and moisture 73.8%. The average vitamin C content for 'Dwarf Brazilian' fruit ranged from 6.3 to 17.5 mg 100 g^{-1}, and 'Williams' ranged from 2.5 to 6.3 mg 100 g^{-1} from four locations in Hawai'i. Ascorbic acid content can decrease during ripening (Wenkam 1990). Ascorbic acid content of 'Dwarf Cavendish', 'Rasabale' and 'Rajabale' increased during ripening at 20 °C for 21 days and then decreased slightly up to 35 days (Desai and Deshpande 1975). In India Deekshika et al. (2015) compared the vitamin C content of mangoes and bananas and found that mangoes had 54.78 ± 2.19 mg 100 g^{-1} and bananas had 20.13 ± 1.54 mg 100 g^{-1}. Wills et al. (1984) detected 1.4 mg dehydroascorbic acid 100 g^{-1} in some of the bananas they tested but not in others. Vanderslice et al. (1990) reported 3.3 mg 100 g^{-1} dehydroascorbic acid in bananas. These results may be influenced by the particular test used.

Table 3.11 Ascorbic acid content of the pulp of ripe bananas. Modified from Seenappa et al. (1986)

Genotype	Ascorbic acid mg 100 g^{-1}
'Dwarf Cavendish' *Musa* AAA	4.4
'Ney Poovam' *Musa* AB	7.9
'Silk' *Musa* AAB	9.3

Table 3.12 Ascorbic acid content of the pulp of unripe bananas. Modified from Seenappa et al. (1986)

Genotype	Ascorbic acid mg 100 g^{-1}
'Bluggoe' *Musa* ABB	16.0
'Robusta' *Musa* AAA	10.3
'Dwarf Cavendish' *Musa* AAA	7.7
'Gros Michel' *Musa* AAA	6.3

Carotenoids and Vitamin A

Vitamin A is crucial for human health, since it plays an important role in vision, cell growth and the regulation of the immune system, but it cannot be synthesised in sufficient quantities by the body and must be obtained from the diet. Plants do not contain vitamin A but they contain carotenoids. Some carotenoids can be converted into vitamin A during human digestion and are called provitamin A carotenoids (pVACs). In meat retinol is the main form of vitamin A. Measurement of pVAC is commonly related to Retinol Activity Equivalent and is used for quantifying the vitamin A value of sources of vitamin A, including both preformed retinoids in animal foods and precursor carotenoids in plant foods. Wall (2006) reported that bananas are generally low in pVACs (Table 3.6), but some Fe'i bananas that have yellow or orange pulp have been shown to contain relatively high levels of carotenoids (Table 3.13). In ripe bananas, the major carotenoids were lutein and carotene. 'Dwarf Brazilian' bananas had 96.9 mg β-carotene and 104.9 mg α-carotene 100 g^{-1}, whereas 'Williams' averaged 55.7 mg β-carotene and 84.0 mg α-carotene 100 g^{-1}. She also reported that bananas contained higher concentrations of lutein than of the pVACs. Average lutein concentrations were 154.9 mg 100 g^{-1} for 'Dwarf Brazilian' and 108.3 mg 100 g^{-1} for 'Williams'. In Hawai'i, 'Apple' bananas had an average of 96.9 µg β-carotene and 104.9 µg α-carotene 100 g^{-1}, whereas 'Williams' bananas averaged 55.7 µg β-carotene and 84.0 µg α-carotene 100 g^{-1}.

Most carotenoids in bananas are in the peel with generally low amounts in the pulp. Seymour (1986) found that the carotenoid content of banana peel could change during ripening, depending on temperature. Those ripened at 35 °C had significantly increased carotenoids in the peel, while for in those ripened at 20 °C carotenoids remained constant. The typical yellow colour of ripe banana peel is developed purely because of the breakdown of chlorophyll, which masks the yellow colour in unripe bananas. At high temperatures, the fruit remained greenish in spite of carotenoid synthesis, because chlorophyll content was only partially reduced. Wenkam (1990) also found that the level of carotenoids increased during maturation and ripening. These results confirm those of Von Loesecke (1949) and contradict those of Gross and Flugel (1982) who found a decrease in carotenoids during the initial phase of ripening of bananas. Measurement of the pVACs in 171 banana genotypes

Table 3.13 Carotenoid content of different Fe'i banana varieties. Modified from Sidhu and Zafar (2018) and Englberger et al. (2006)	Variety	Total carotenoid content µg 100 g^{-1} fresh weight
	'Aibwo', 'Suria'	9400
	'Fagufagu'	5054
	'Ropa'	5218
	'Gatagata'	774
	'Toraka Parao'	776

showed that levels varied from undetectable in some fruit with white-cream pulp to 3500 μg 100 g^{-1} for fruit with yellow or orange pulp (Davey et al. 2009, Pereira et al. 2011). Among Indian banana varieties Arora et al. (2008) reported 'Red Banana' ranked highest in total carotenoids content for pulp (4 μg g^{-1} dry weight) and β-carotene was estimated to be the highest in the peel (241.91 μg 100 g^{-1}) and in pulp (117.2 μg 100 g^{-1}). Wall (2006) reported that 'Williams', harvested from four locations in Hawai'i, had 8.2 mg RAE 100 g^{-1} for vitamin A and ranged from 6.1 to 9.3 mg RAE 100 g^{-1}. This was higher than the 4.5 mg RAE 100 g^{-1} reported by Wenkam (1990) for 'Cavendish'. Sidhu and Zafar (2018) also showed that there was considerable variation in carotenoids content in different banana genotypes (Table 3.13 and Table 3.14).

The distribution of specific carotenoids in 'Cavendish' from the West Indies, Central and South American were identified by Seymour et al. (1987) and were predominantly lutein (Table 3.15). They also showed that levels of the different carotenoids they detected could change during ripening and this could be affected by the ripening temperature, but no consistent trends in these changes could be detected.

Table 3.14 Carotenoid content of different banana varieties. Modified from Sidhu and Zafar 2018 and Beatrice et al. (2015)

Variety	Total carotenoid content μg 100 g^{-1} fresh weight
'Hung Tu' *Musa* AAA	7,760
'To'o' *Musa* AA	7,765
'Sepi' *Musa* AA	10,067
'Apantu', 'False Horn' plantain *Musa* AAB	10,056
'Bungaoisan', 'Ntanga 4' *Musa* AAB	1,675

Table 3.15 Major carotenoid content in the peel of 'Cavendish' bananas before being initiated to ripen and after being initiated to ripen with 1000 μL L-1 ethylene for 24 hours and ripened at either 20 or 34 °C for 5 days. Carotenoids were separated by HPLC with acetonitrile as solvent and identified by comparison with authentic standards. Modified from Seymour et al. (1987)

Carotenoid	Percentage total carotenoid content		
	Unripe fruit	Ripened at 20 °C	Ripened at 34 °C
Lutein	42.9	45.3	36.3
α and β carotene	7.1	9.8	7.3
Neoxanthin	3.8	2.6	3.2
Violaxanthin	3.5	3.3	6.8
Unknown	3.2	3.7	1.6
Unknown	1.2	10.0	7.2
Unknown	4.7	2.3	3.8
Unknown	0.3	3.0	1.7

Folates

Folates are vitamin B_9 and are a group of compounds derived from pteroylglutamic acid. They are essential for in human development including the prevention of neural tube defects and cardiovascular and neurodegenerative diseases. Folate levels in bananas was given as 20 µg DFE (1 microgram dietary folate equivalent = 0.6 µg folic acid) (Table 3.6). García-Salinas et al. (2016) showed fluctuations in total folate in bananas throughout ripening, but levels returned to their initial mature green values when they were fully ripe. Interestingly, they found that initiating ripening with exogenous ethylene increased folate by 51 %.

Other Phytochemicals

A phytochemical is a natural bioactive compound found in plant food that works with nutrients and dietary fibre to protect against disease. Imam and Akter (2011) in their phytochemical and pharmacological review gave the following phytochemicals in various parts of bananas and plantains:

"catecholamines such as norepinephrine, serotonin, dopamine,
tryptophan, indole compounds,
pectin in the pulp,
flavonoids and related compounds (leucocyanidin, quercetin and its 3-Ogalactoside, 3-O-glucoside, and 3-O-rhamnosyl glucoside) from the unripe pulp of plantain.
serotonin, nor-epinephrine, tryptophan, indole compounds, tannin, starch, iron, crystallisable and non-crystallisable sugars, vitamin C, B-vitamins, albuminoids, fats, mineral salts in the fruit pulp.
acyl steryl glycosides such as sitoindoside-I, sitoindoside-II, sitoindoside-III, sitoindoside-IV and steryl glycosides such as sitosterol gentiobioside, sitosterol *myo*-inosityl- β-D-glucoside.
a   bicyclic   diarylheptanoid,*rel*-(3*S*,4a*R*,10b*R*)-8-hydroxy-3-(4-hydroxyphenyl)-9-methoxy-4a,5,6,10b-tetrahydro-3*H*-naphtho[2,1-*b*]pyran,   and 1,2-dihydro-1,2,3-trihydroxy-9-(4-methoxyphenyl)phenalene, hydroxyanigorufone, 2-(4-hydroxyphenyl)naphthalic anhydride, 1,7-bis(4-hydroxyphenyl)hepta-4(*E*),6(*E*)-dien-3-one.
several triterpenes such as cyclomusalenol, cyclomusalenone, 24- methylenecycloartanol, stigmast-7-methylenecycloartanol, stigmast −7-en-3-ol, lanosterol and β-amyrin.
an antihypertensive principle, 7, 8-dihydroxy-3-methylisochroman-4-one from the fruit peel.
cycloartane triterpenes such as 3-epicycloeucalenol, 3-epicyclomusalenol, 24-methylenepollinastanone, 28-norcyclomusalenone, 24-oxo-29- norcycloartanone have been isolated from the fruit peel.

cellulose, hemicelluloses, arginine, aspartic acid, glutamic acid, leucine, valine, phenylalanine and threonine have been isolated from pulp and peel.

hemiterpenoid glucoside (1,1-dimethylallyl alcohol), syringin, (6S, 9R)-roseoside, benzyl alcohol glucoside, (24R)-4α,l4 α,24-trimethyl-Sacholesta-8,25(27)-dien-3β-ol have been isolated from the flower."

The ripe pulp of 'Chinese Cavendish', 'Giant Cavendish', 'Dwarf Red', 'Grand Nain', 'Eilon', 'Gruesa', 'Silver', 'Ricasa', 'Williams' and 'Zelig' had similar amounts of lipophilic extracts (about 0.4 % of DW). The major groups of lipophilic components identified in these fractions were fatty acids (68.6–84.3 %) and sterols (11.1–28.0 %), with smaller amounts of long chain aliphatic alcohols and α-tocopherol (Vilela et al. 2014). Chanai Noysang et al. (2019) measured the phytochemicals in ethanolic extracts of unripe pulp and peel of 'Kluai Namwa', 'Kluai Hom Thong', 'Kluai Leb Mu Nang' and 'Kluai Khai' bananas. Kluai Khai pulp had the highest amount of ethanolic extractives and the main phytochemicals identified were alkaloids, flavonoids, tannins and polyphenols. 'Kluai Hom Thong' peel ethanolic extract had the highest antioxidant activity.

Chapter 4
Postharvest Treatments to Control Ripening

Controlled Atmosphere Storage

CA on Pre-Climacteric Bananas

Burg and Burg (1967) showed that plant tissue was generally less sensitive to ethylene in atmospheres containing lower O_2 or higher CO_2 levels (Table 4.1). As described earlier ethylene biosynthesis initiates banana ripening and the pre-climacteric period of 'Cavendish' bananas was shortened by exposure to levels of ethylene as low as 0.1 ppm, with greater sensitivity to ethylene the more mature the fruits at harvest (Liu 1976a), but less sensitive to ethylene in reduced O_2 (4 kPa) and increased CO_2 (7 kPa) concentrations (Liu 1976b). Ahmad and Thompson (2007) observed that the effectiveness of ethylene in hastening ripening was not reduced with CO_2 levels of less than 5 kPa and O_2 levels of greater than 10 kPa. Previously, Quazi and Freebairn (1970) showed that increased CO_2 had the effect of inhibiting the synthesis of endogenous ethylene and the respiration rate of bananas when the pulp ripened. In an atmosphere containing 1 kPa O_2, ripening was not initiated even when exogenous ethylene was applied.

The production of ethylene was inhibited when the concentration of atmospheric O_2 was lowered to about 7.5 kPa and below, and tissues were then insensitive to ethylene. Quazi and Freebairn (1970) showed that high CO_2 and low O_2 delayed the production of ethylene in pre-climacteric bananas, but the application of exogenous ethylene was shown to reverse this effect. Truter and Combrink (1990) found that 'Dwarf Cavendish' and 'Williams' remained hard and green for 6 weeks under 2 kPa O_2 and 7 kPa CO_2 at 12.5 °C. Thereafter the fruit ripened normally within 4–5 days when transferred to normal atmosphere at 16 °C. Ahmad et al. (2006) studied the effects of CA storage (2 kPa O_2 combined with 4, 6 or 8 kPa CO_2) for 2 weeks on ripening of 'Cavendish' after removal to air. All treatments responded to

© The Author(s), under exclusive license to Springer Nature Switzerland AG 2019
A. K. Thompson et al., *Banana Ripening*, SpringerBriefs in Food, Health, and Nutrition, https://doi.org/10.1007/978-3-030-27739-0_4

Table 4.1 Sensitivity at 20 °C of pea seedlings to ethylene in different levels of O_2, and CO_2 with the balance being N_2 (Burg and Burg 1967)

O_2 kPa	CO_2 kPa	Sensitivity to ethylene $\mu L\ L^{-1}$
0.7	0	0.6
2.2	0	0.3
18.0	0	0.14
18.0	1.8	0.3
18.0	7.1	0.6

ripening initiation with ethylene, in a similar way to fruit that had not been exposed to CA, and produced good quality ripe fruit. All bananas, including the control, reached colour stage 6 (Table 1.1) in 9 days after ethylene treatment. Imahori et al. (2013) stored mature green 'Cavendish' in 0.5, 2, or 21 kPa O_2 for 7 days at 20 °C before ripening was initiated with ethylene. Concentrations of ethanol in mature green fruit did not change during storage in both 21 and 2 kPa O_2 atmospheres, but increased in fruit stored in 0.5 kPa O_2. They concluded that 0.5 kPa O_2 is too low for CA storage of pre-climacteric bananas but 2 kPa O_2 was suitable.

At 15.5 °C ripening of bananas was retarded when they were stored in total N_2, but their flavour was poorer only if they were held in these conditions for longer than 4 days or for over 10 days in 99 kPa N_2 + 1 kPa O_2 (Ahmad et al. (2006). Klieber et al. (2002) found that storage of bananas in total N_2 at 22 °C did not extent their storage life compared to those stored in air but resulted in brown discolouration.

The effects of CA storage on banana quality was tested by Kanellis et al. (1993) who found that if CO_2 concentrations exceeded 10 kPa, the pre-climacteric period was prolonged, but the organoleptic qualities and the peel colour of 'Pisang Mas' was detrimentally affected. However, the quality of the 'Lacatan' (*Musa* AAA) and 'Latundan' (*Musa* AAB) stored in PE film bags under an atmosphere of 5 kPa O_2 and 12.5 kPa CO_2 at a temperature of 26–30 °C, was reported not to affect their quality (Tiangco et al. 1987). Kanellis et al. (1993) have described how storage in 2.5 kPa O_2 reduced respiration rate, slowed sugar accumulation and arrested the normal colour changes in bananas. They concluded that the observed effects were mediated via ethylene action rather than as a direct effect on respiration.

CA Effects After Ripening Initiation

Hesselman and Freebairn (1969) showed that ripening of bananas, which had already been initiated to ripen by exogenous ethylene, was slowed in a low O_2 atmosphere. Wade (1974) showed that bananas could be ripened in atmospheres of reduced O_2, even as low as 1 kPa, but the peel failed to degreen, which resulted in ripe fruit which were still green. Similar effects were shown at O_2 levels as high as 15 kPa. Since the degreening process in 'Cavendish' is entirely due to chlorophyll degradation (Seymour et al. 1987a) the controlled atmospheres storage effect was presumably due to suppression of this process. Maneenuam and Doorn (2007) showed that 'Sucrier' that had been initiated to ripen and then exposed to 90 ± 2 kPa O_2 had significantly increased peel spotting and the activity of PAL and PPO were lower compared to the controls (18 ± 2 kPa O_2). Quazi and Freebairn (1970) showed that high CO_2 and low O_2 delayed the increased production of ethylene associated with the initiation of ripening in bananas, but the application of exogenous ethylene was shown to reverse this effect. Wade (1974) showed that bananas could be ripened in atmospheres of reduced O_2, even as low as 1 %, but the peel failed to degreen, which resulted in ripe fruit which were still green. Similar effects were shown at O_2 levels as high as 15 %. Since the degreening process in Cavendish bananas is entirely due to chlorophyll degradation (Seymour et al. 1987, Blackbourn et al. 1990), the controlled atmosphere storage treatment was presumably due to suppression of this process. Hesselman and Freebairn (1969) showed that ripening of bananas, which had already been initiated to ripen by ethylene, was slowed in low O_2 atmospheres.

Hypobaric Storage

A considerable amount of work has been done on the effects of reduced pressure in the storage atmosphere of bananas by Stanley Burg. In his review Burg (2004) reported that at 13.3–14.4 °C 'Valery' and 'Gros Michel' ripened in 10 days under atmospheric pressure but remained green for 40–50 days under 152 mm Hg, but mould could develop during this protracted period, which could damaage the fruit. Burg (2004) also reported that 'Valery' stored at 13.3 °C under 6.4, 9.5 or 12.7 kPa remained green for more than 105 days and ripened normally when they were transferred to atmospheric pressure and exposed to exogenous ethylene, with no deleterious effects on flavour or aroma. Burg (2004) also stored bananas under 49, 61, 76, 101, 122 or 167 mm Hg for 30 days and found they lost 1.1–3.6% in weight with the higher losses at the higher pressures because of their higher respiration rates. All the fruit were still green and there were no apparent differences between the different hypobaric conditions. Burg and Burg (1965) showed delays in ripening of bananas when they were stored under 125–360 mm Hg compared to those under atmospheric pressure.

Apelbaum et al. (1977) stored 'Dwarf Cavendish' at 14 °C under 81, 253 or 760 mm Hg pressure and found that they began to turn yellow after 30 days under atmospheric pressure, 60 days under 253 mm Hg but they still remained dark green under 81 mm Hg. When they were subsequently transferred to atmospheric pressure at 20 °C and exposed to exogenous ethylene they all ripened to a good flavour, texture and aroma. Bangerth (1984) successfully stored 'Cavendish' at 14 °C under 51 mm Hg for 12 weeks and found that when they were subsequently ripened in 50 µl L^{-1} exogenous ethylene they were the same quality as freshly harvested fruit or those that had been stored under atmospheric pressure.

Research has also been done on the effects of reduced pressure storage on the response of bananas to exposure to exogenous ethylene. Bangerth (1984) found that there was no diminished response to ethylene in storage under 51 mm Hg, whatever the storage temperature tested. Liu (1976c) stated that at 14 °C the duration of the pre-climacteric period of bananas, even those that had been exposed to ethylene, was prolonged under low pressure (0.1 atmospheric pressure) or at a low concentration of O_2. As well as delaying the initiation of ripening, hypobaric storage has also been shown to delay the speed of ripening of bananas that have been initiated to ripen. Quazi and Freebairn (1970) found that bananas that had been initiated to ripen by exposure to exogenous ethylene did not ripen during hypobaric storage but began to ripen within 1–2 days after transfer to ambient atmospheric pressure and had good eating quality and texture. Liu (1976) initiated 'Dwarf Cavendish' to ripen by exposure to 10 μL L^{-1} ethylene at 21 °C and then stored them at 14 °C for 28 days under hypobaric storage of 51 and 79 mm Hg or controlled atmosphere storage of 1% O_2 + 99% N_2. All fruit remained green and firm and continued to ripen normally after they had been removed to ripening temperature in atmospheric pressure.

Modified Atmosphere Packing

MAP has been used successfully to slow the ripening of bananas after they have been initiated to ripen and to delay their ripening initiation (Fig. 4.1). This effect is primarily due to changes in the O_2 and CO_2 in and around the fruit but, as reported by Thompson et al. (1974), reducing fruit weight loss can also contribute to this effect. This can also be shown with the effect of moist coir and perforated PE bags that substantially delayed ripening, but were less effective that non-perforated PE bags that would both reduce weight loss and substantially change the O_2 and CO_2 around the fruit (Table 4.2).

Nair and Tung (1988) reported that 'Pisang Mas' had an extension in their post-harvest life of 4–6 weeks at 17 °C when they were stored in evacuated collapsed PE film bags by applying a vacuum not exceeding 300 mm Hg. There are many other reports of the positive effects of MAP on different varieties of banana in different films from different countries including: Shorter et al. (1987), Tiangco et al. (1987), Tongdee (1988), Abdullah and Pantastico (1990), Marchal and Nolin (1990), Satyan et al. (1992), Wei and Thompson (1993), Chamara et al. (2000) and Chauhan et al. (2006).

Vacuum packing is where bananas are placed in a plastic bag and the air is extracted with a vacuum pump and the bags are sealed. MAP is commonly used for the international banana trade (Fig. 4.2). Banavac® is a patented system which uses large 0.04 mm thick PE film bags in which typically 18.14 kg of green bananas are packed then a vacuum is applied and the bags are sealed and transported in cartons (Badran 1969). Nair and Tung (1988) reported that 'Pisang Mas' bananas had an extension in their postharvest life of 4–6 weeks at 17 °C when they were stored in

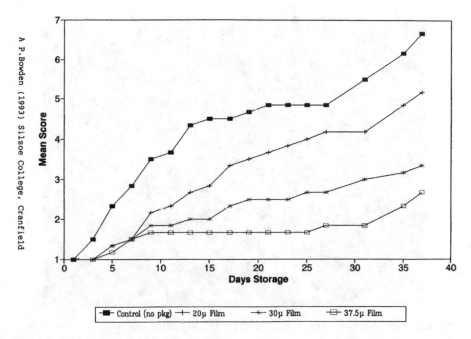

Fig. 4.1 Changes in skin colour (1 = dark green and 6 = fully yellow) of five finger clusters of 'Apple' bananas stored at 13 °C in different thickness of polyethylene film bags after being initiated to ripen. Source: Bowden 1993

Table 4.2 Effects of wrapping and packing material on time to ripening and weight loss at Ripening Index 7 of plantains stored at tropical ambient conditions of 26–34 °C and 52–87 % RH (modified from Thompson et al. 1972)

Packing material	Days to ripeness	Weight loss at ripeness
Not wrapped	15.8	17.0 %
Paper	18.9	17.9 %
Moist coir fibre	27.2	(3.5 %)[a]
Perforated PE	26.5	7.0 %
PE	36.1	2.6 %
LSD ($p = 0.05$)	7.28	2.81

[a]The fruit actually gained 3.5% in weight

evacuated collapsed polyethylene film bags by applying a vacuum not exceeding 300 mm Hg.

Ahmad and Thompson (2007) tested the effect of MAP on the ripening and quality of ripe fruit. Bananas ripened in 0.05 mm of PE film bags had a shelf-life extended by 5 days, as well as a slower the rate of softening and a more attractive fresh appearance than those not in plastic bags. Organoleptic panellists preferred the bananas ripened in PE bags to those not in PE bags because of their better flavour and appearance. Bowden (1993) showed that the effect of MAP on ripening (changes in yellowing of the peel) of bananas that had been initiated to ripen was related to

Fig. 4.2. 'Gros Michel' vacuum packed in polyethylene film ready for export

the thickness of plastic and therefore the gas composition within the film (Fig. 4.1). Yao et al. (2014) stored plantains sealed in polyethylene film bags of different thicknesses at 28 ± 2 °C and found that the concentration of O_2 was reduced to 3.5 kPa after 14 days in 20 and 30 μ bags, to 0.5 kPa in 40, 50 and 60 μ bags and to 0.3 kPa in 70 and 80 μ thick. Film thickness also affected weight loss, for example, at 13 °C Wei and Thompson (1993) showed that weight loss of 'Apple' bananas after 4 weeks storage in PE film was 1.5% in 50 μm, 1.8 % in 37.5 μm and 2.1 % in 25 μm while for fruit stored without packaging it was 12.2 %.

Ethylene Absorbents

Ethylene absorbents have been used, usually inside MAP to extend the postharvest life of many types of fruit. Various brands have been marketed commercially e.g. Green Keeper (GK) and Ethysorb, which contain potassium permanganate. More recently sachets containing palladium (It's Fresh!) has been successfully used as ethylene absorbents. Kulkarni et al. (2011) stored 'Robusta' under active and passive modified atmosphere packaging at 12 ± 1 °C and 85–90 % RH for 2 seasons and found that after 3 weeks a steady state of about 8.6 and 8.2% of CO_2 and 2.8 and 2.6 % of O_2 in passive MAP and MAP+GK packages, respectively were established (Fig. 4.3). GK is a brand of sachet containing potassium permanganate. Weight loss was 0.8 % after 7 weeks of storage, as against 5 % in openly kept green bananas after 3 weeks. Both MAP and MAP+GK treatments delayed colour, texture, pulp to

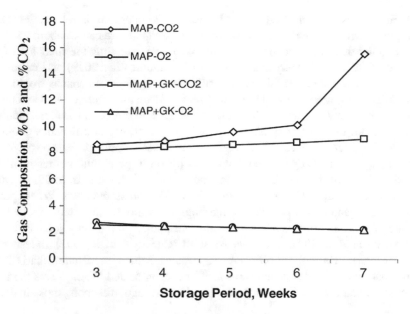

Fig. 4.3 Gas composition of MAP and MAP+GK for 'Robusta' in LDPE films. GK = KMnO$_4$ sachets as Green Keeper. Modified from Kulkarni et al. (2011)

peel ratio and TSS content compared to controls that were stored non-wrapped. Results indicated that the shelf-life of fruits packed under MAP and MAP+GK could be extended by up to 5 and 7 weeks, respectively as compared to 3 weeks for openly kept control fruits. Sensory quality of fully ripe fruits of both passive MAP and MAP+GK treatments after ripening was very good.

Ethylene absorbents have also been used in banana transport. Scott et al. (1971) found that during transport an alternative to refrigeration was sealing hands of green fruit in polythene bags with an ethylene absorbent (potassium permanganate). The fruit remained in a green, firm condition for up to 18 days at ambient temperatures, during transport from North Queensland in Australia to Auckland in New Zealand. This technique has also been successfully demonstrated to delay the ripening of whole bunches of bananas during shipment (Wills et al. 1998). They estimated that about 2 weeks additional storage life was obtained by packing potassium permanganate in sealed bags with the fruit. Chauhan et al. (2006) stored unripe 'Pachbale' bananas at 13 ± 1 °C in modified atmospheres consisted of PE film pouches followed by Ethrel induced ripening at 30 ± 1 °C. The PE packaging used was: not flushed or flushed with 3 % O$_2$ and 5 % CO$_2$ or partial vacuum (400 mmHg) giving a shelf-life extension to 15, 24 and 32 days, respectively as against 12 days for the control. Application of an ethylene scrubber in combination with silica gel as a desiccant and soda-lime as a CO$_2$ scrubber further enhanced the shelf-life to 18, 28 and 36 days, respectively under the different types of modified atmospheres specified.

One supermarket in Thailand used ethylene absorbing sachets in their MAP bananas in an attempt to increase their shelf-life both whilst being offered for sale on their shelves and for customers at home. The temperature is relatively high in both places and even under air-conditioning the temperature can be about 25 °C. It was reported that this treatment was unsuccessful in extending the shelf-life of the bananas (Patcharin Chitaurjaisuk personnel communication 2019) and it was not used. It was suggested that this was not surprising since these bananas would have been initiated to ripening by ethylene treatment before they were packed and therefore absorbing any ethylene in the pack would have no effect on speed of ripening.

Scott ct al. (1971) found the storage life of pre-climacteric bananas could be extended by at least 2 weeks by packaging them in sealed polyethylene bags with an ethylene absorbent. They showed that the inclusion of potassium permanganate in sealed packages of bananas reduced the mean level of ethylene from 395 down to 1.5 μL L^{-1} and reduced brown heart from 68 to 36 % in stored pears. Where potassium permanganate was included in the bags containing bananas the increase in storage life was 3–4 times compared to non-wrapped fruit and they could be stored for 6 weeks at 20 or 28 °C and 16 weeks at 13 °C (Satyan et al. 1992). In tests with pre-climacteric 'Harichhaal' (*Musa* AAA) bananas Kumar and Brahmachari (2005) found that potassium permanganate-soaked paper in sealed PE bags was the most effective packaging treatment for extending their storage life among those that they tested.

Chemicals

Metal Ions

Domínguez et al. (1998) showed that exposing bananas to cobalt ions or silver ions inhibited ethylene biosynthesis and both were found to inhibit both ethylene production and ripening, even though silver is believed to act directly on ethylene action only and cobalt ions inhibited induced respiration and ethylene production. Silver ions effectively inhibited both the initiation and the continuation of tomato ripening (Davies et al. 2006). They showed that the application of silver thiosulphate to mature green tomatoes prevented the appearance of several novel proteins associated with ripening, including polygalacturonase.

1-Methylcyclopropene

1-MCP is an ethylene antagonist that has been used for controlling ripening and senescence of fruit and vegetables by inhibiting both ethylene biosynthesis and action by irreversibly binding to ethylene-receptors and strongly inhibiting the autocatalytic production of ethylene (Lurie 2008). Pathak et al. (2003) reported that 1-MCP application directly affected ACS transcript accumulation and inhibited the autocatalytic pathway (System 2) in 'Dwarf Cavendish' bananas but did not affect

System 1 ethylene production. Kesar (2010) found that exposure of unripe mature bananas to exogenous ethylene induced the expression of the gene *MaPR1a*, and 1-MCP exposure prior to ethylene treatment inhibited its expression. Ruiling Liu et al. (2016) suggested that the mode of action of 1-MCP was related to the expression of acid transport genes, including *MdVHA-A*, *MdVHP* and *Ma1*. This was because 1-MCP maintained fruit acidity in apples by regulating the balance between malate biosynthesis and degradation. In addition, Trivedi and Nath (2004) addressed the effects of 1-MCP on fruit firmness maintenance and found that treated bananas had low expression of *Maexp1*, an ethylene induced expansin gene. 1-MCP application also lowered the solubilisation of pectin due to suppression of cell wall hydrolases such as pectin methylesterase, PG, cellulase and pectate lyase activities in bananas (Lohani et al. 2004).

Although 1-MCP treatment can delay the ripening process in climacteric fruit, including banana, the efficiency of 1-MCP treatment is limited according to the maturity of fruit at harvest, 1-MCP concentration, temperature and durations of exposure (Bagnato et al. 2003; Pelayo et al. 2003; Moradinezhad et al. 2008). Harris et al. (2000) suggested that the efficiency of 1-MCP in prolonging shelf-life of bananas varied according to fruit maturity at harvest and in bananas it was less effective at advanced maturity at harvest (Moradinezhad et al. 2008). They also found that 1-MCP treatment was more effective in the fruit in the early part of the climacteric and in hands from the top of bunch rather than from the bottom of the bunch.

Bagnato et al. (2003) reported that the shelf-life of ethylene treated 'Cavendish' bananas was delayed by 300 µL L^{-1} 1-MCP treatment which it was longer than non-treated fruit for 3 days. Jiang et al. (1999a and b) showed that banana sealed in 0.03 mm thick PE film bags containing 1-MCP at either 0.5 or 1.0 µL L^{-1} delayed the ripening by about 58 days compared to non-wrapped, non-treated fruit. The analysis of ethylene and respiration rate within the PE bags confirmed that 1-MCP supressed both ethylene biosynthesis and respiratory rate in the banana fruit. Subsequently Jiang et al. (2004) showed that exposure of bananas to mL L^{-1} 1-MCP slowed ripening but also enhanced chilling injury that was associated with increased membrane permeability. Pinheiro et al. (2005) had studied the effect of 1-MCP concentration on postharvest quality of 'Apple' bananas during ripening. They found that 1-MCP treatment for 12 h at the concentration of 50 nL L^{-1} was the best concentration in extending shelf-life based on total sugar, soluble pectins, firmness and external appearance at the end of storage compared to higher concentrations of 1-MCP. In 'Kluai Khai' bananas (*Musa* AA), 1-MCP treatment at 250 ppb for 24 h lowered respiratory rate and ethylene evolution as well as delaying ripening and also prolonged their shelf-life to some 20 days during storage at 20 °C compared to the fruit treated with lower concentrations (Jansasithorn and Kanlayanarat 2006).

Kesar (2010) found that exposure of unripe mature bananas to exogenous ethylene induced the expression of the gene *MaPR1a*, which increased with ripening and 1-MCP treatment prior to ethylene exposure, thus inhibiting gene expression. Ruiling Liu et al. (2016) suggested that the mode of action of 1-MCP was related to the expression of acid transport genes, including *MdVHA-A*, *MdVHP* and *Ma1* because 1-MCP maintained fruit acidity by regulating the balance between malate biosynthesis and degradation. 1-MCP delayed the postharvest loss of malate and citrate.

With 'Cavendish' bananas harvested at the mature-green stage, exposure to $0.01-1.0$ µL L^{-1} 1-MCP for 24 h delayed peel colour change and fruit softening, extended shelf-life and reduced respiration rate and ethylene production (Jiang et al. 1999). They obtained similar results with bananas sealed in 0.03 mm thick PE bags containing 1-MCP at either 0.5 or 1.0 µL L^{-1}, but delays in ripening were longer at about 58 days. Analyses of ethylene and CO_2 concentrations within the PE bags confirmed that 1-MCP suppressed both ethylene evolution and respiration rate.

Bananas treated with 1-MCP were shown to result in uneven degreening of the peel. De Martino et al. (2007) initiated 'Williams' to ripen by exposure to 200 µL L^{-1} of ethylene for 24 h at 20 °C. These bananas were then treated with 200 nL L^{-1} 1-MCP at 20 °C for 24 h then stored in > 99.9 kPa N_2 + < 0.1 kPa O_2 in perforated PE bags at 20 °C. No differences in the accumulation of acetaldehyde and ethanol were detected during storage compared to fruits not treated with 1-MCP. Peel degreening, the decrease in chlorophyll content and chlorophyll fluorescence, were delayed after the 1-MCP treatment. There was some general browning throughout the 1-MCP treated peel in both the green and yellow areas of the ripening peel. They concluded that it appears the 1-MCP treated peel, 24 h after the ethylene treatment, may still undertake some normal senescence that occurs during banana ripening. Green mature 'Apple' bananas were treated with 1-MCP at 0, 50, 100, 150 or 200 nL L^{-1} for 12 h and then stored at room temperature of 20 ± 1 °C and 80 ± 5 % RH. 50 nL L^{-1} 1-MCP was the best concentration in extending the shelf-life based on total sugars, soluble pectins, firmness and external appearance in the end of storage. The 50 nL L^{-1} 1-MCP delayed the onset of peel colour change by 8 days while at 100, 150 and 200 nL L^{-1} 1-MCP the delay was by 10 days compared to non-treated (Pinheiro et al. 2005). 'Khai' bananas were treated with 1-MCP at 50, 100 or 250 ppb for 24 h then stored at 20 °C. 1-MCP delayed ripening and prolonged storage to some 20 days with the higher concentration being more effective, but bananas treated with 250 ppb 1-MCP had the lowest respiration rate and ethylene production. However, there is no significant difference among treatments for colour changes (Jansasithorn and Kanlavanarat 2006). Joyce et al. (1999) found that banana ripening induced by propylene, could be delayed by exposure to 15 ml^{-1} L^{-1} of 1-MCP at 20 8 °C for 12 h. However, the 1-MCP treatment was less effective as propylene-induced ripening progressed, although the eating-ripe condition of fruits was maintained for a longer time than the non-treatment. Similarly, Yueming et al. (1999) found that 1-MCP applied at concentrations in the range of $0.01-10$ ml^{-1} L^{-1} at 20 °C for 12 h 1 day after ethylene treatment slowed the ripening of bananas, but it was ineffective when applied 3 or 5 days after ethylene treatment.

1-MCP delayed or slowed ripening of pre-climacteric and climacteric bananas. 1-MCP treated pre-climacteric fruit eventually ripened, possibly due to synthesis of new ethylene receptors. 1-MCP treatment became less effective as ripening progressed, especially after peak ethylene production. Nevertheless, 1-MCP treatment extended the shelf-life of fruit when applied at later ripening stages, which may be a commercial advantage. Thus, 1-MCP promises improved control of banana fruit ripening when applied before or after exposure to exogenous ethylene (Joyce et al. 1999). Deaquiz et al. (2014) tested the interaction between 1-MCP (600 mg L^{-1}) and Ethrel (3 mL L^{-1}) on ripening of yellow pitahaya fruits (*Selenicereus megalanthus*) and concluded that the 1-MCP reduced ethylene action and slowed the ripening. A patented system, marketed as RipeLock™ technology, treats bananas, after they have been initiated to ripen, with 1-MCP and packs them in micro-perforated film. The company claim that this combination of treatments "extends banana yellow life 4–6 days over any existing technologies." (Anonymous 2019).

Salicylic Acid

Salicylic acid is a plant growth regulator that has been shown to stimulate defence mechanism against both biotic and abiotic stresses and can inhibit ethylene biosynthesis. It is widely accepted as a postharvest treatment for fruit and vegetables (Asghari and Aghdam 2010; Supapvanich and Promyou 2013). Its mode of action was reported to induce the activity of phenylananine ammonia lyase (PAL) activity triggering the phenylpropanoid pathway resulting in the accumulation of phenolic compounds and antioxidant activity in plants. It also induces heat shock proteins and retards lipid peroxidation of membrane leading to increased tolerance to chilling temperatures (Supapvanich and Promyou 2013; Aghdam et al. 2014). Pan et al. (2007) also reported that treatment with salicylic acid alleviated chilling injury symptoms by retarding relative electrical conductivity of tissue during storage of bananas. They also found that salicylic acid retarded respiration rate, the change of banana peel colour and the increases in PPO activity during storage. Thus, postharvest treatment with salicylic acid has been used to control fruit ripening and physiological disorders caused by stress during storage. Srivastava and Dwivedi (2000) reported that postharvest immersion of 'Harichhal' bananas in salicylic acid at 500 or 1000 μM for 6 h could delayed their ripening. Fruit softening (caused by major cell wall hydrolases such as PG, xylanase and cellulose), pulp to peel ratio, reducing sugars content, invertase activity, respiration rate and the major antioxidant enzymes (catalase and peroxidase) were all lower in salicylic acid treated bananas compared to control. Anuchai et al. (2018) also showed that immersion for 30 mins in 2 mM salicylic acid could maintain visual appearance and delayed senescence of 'Hom Thong' bananas.

Gibberellic Acid

Vendrell (1970) reported that postharvest immersion in aqueous solution of GA_3 at 10^{-6} to 10^{-2} M could delay ripening and yellowing of the peel of bananas. However, Rossetto et al. (2003) found that exogenous application of GA_3 to bananas impaired the onset of starch to sugars conversion and delayed sucrose accumulation by at least 2 days, which they attributed to the disturbance of sucrose phosphate synthase activity but GA_3 application did not affect their ethylene synthesis or respiratory rate. Archana and Sivachandiran (2015) reported that GA_3 immersion of 'Kathali' bananas at 500 or 750 ppm prolonged their storage life, delayed peel colour change and sugar accumulation, reduced the rate of increasing pH and the loss of fresh weight.

Diazocyciopentadiene

The application of DACP to tomato fruit, while still on the plant, was shown to retard the development of red color by blocking the ethylene receptor sites (Sisler and Blankenship 1993). DACP is a weak inhibitor of ethylene responses, but upon irradiation with visible light gives rise to much more active components (Sisler 1991), but it is prohibited for commercial use (Sisler and Lallu 1994).

Indole-3-Acetic Acid

Auxins are plant hormones that can retard ripening and senescence of fruit due to suppression of ethylene. The effect of IAA on banana ripening is associated with the reduction of the climacteric rise in respiration rate and starch to sucrose formation possibly by affecting the activity of β-amylase (Purgatto et al. 2001; 2002). Lohani et al. (2004) found that pre-treating 'Dwarf Cavendish' bananas with either 1-MCP or IAA, suppressed the effect of exogenous ethylene application on the activities of cell wall hydrolases such as pectin methylesterase, PG, pectate lyase and cellulose over the period of 7 days after ripening was initiated by ethylene. Fernández-Falcón et al. (2003) reported that spraying 'Dwarf Cavendish' bananas in the field with IAA could induce their resistance to Panama disease.

Abscisic Acid

ABA is a plant hormone regulating fruit ripening and senescence in both climacteric and non-climacteric fruits (Leng et al. 2014). In climacteric fruit, endogenous ABA levels increased before ripening initiation and then decreased as the fruit ripened. In both tomatoes and bananas there was an increase in ABA prior to the increase in ethylene synthesis during ripening, and exogenous application of ABA induced ethylene production through biosynthesis gene expression (Jiang et al. 2000; Zhang et al. 2009a). However, in non-climacteric fruit, ABA levels increased from maturation to harvest (Setha 2012; Leng et al. 2014). Treatment of bananas with ABA by vacuum infiltration increased respiration rate and hastened ripening, effects that were apparently independent of ethylene. Suppression of ABA led to a delay in fruit ripening (Sun et al. 2012). Lohani et al. (2004) showed that in bananas ABA can act synergistically with ethylene to promote softening. Pre-treating 'Dwarf Cavendish' bananas with ABA stimulated activities of PME, PG, pectate lyase and cellulose and was most evident for pectate lyase. So, ABA has been shown to be able to increase softening of the bananas with or without ethylene. Moreover, ABA, like indole acetic acid, induces an increase in enzyme activity such as that of catalase, peroxidase, acid phosphatase, phenoloxidases, unlike gibberellins (Vendrell 1969).

Lysophosphatidylethanolamine

LPE, a natural glycerolipid, has been approved for use as plant growth regulator for horticultural application in some countries. It was first patented in 1992 (US 5126155) where it is claimed to enhance fruit ripening and delay their senescence (Amaro and Almeida 2013). Postharvest treatment with LPE has been applied to a range of fruits such as tomato (Farag and Palta 1993), red pepper (Kang et al. 2003), cranberries (Ozgen et al. 2005) and Thompson seedless grapes (Hong et al. 2007). In banana fruit, Ahmed and Palta (2015) reported that postharvest dip treatment with LPE could increase marketability of 'Cavendish' bananas by improving shelf-life of the fruit by 1–2 days. The ripening and senescence of the LPE treated bananas were delayed due to the reduction of respiration rate and also by maintaining membrane integrity and decreasing in breakdown of starch and cell walls. To improve water solubility of LPE, the combination of LPE and lecithin was applied to bananas by Ahmed and Palta (2016). They found that the combination of 200 mg L^{-1} LPE and lecithin could increase marketability of the fruit for 7 days after treatment and gave better quality, involving higher pulp firmness, lower ion leakage from peel and thicker peel as compared to LPE alone. Improved membrane integrity with a reduction in electrolyte leakage and ethylene biosynthesis was also observed in banana peel treated with LPE (Ahmed and Palta 2011).

Aminoethoxy-Vinylglycine

AVG is an ethylene antagonist inhibiting the conversion of methionine to ACC (Yang et al. 1979) and was shown to retard ethylene biosynthesis in bananas through inhibiting ACC content and ACS activity (Cuadrado et al. 2008). Domínguez et al. (1998) showed that banana slices, which had been infiltrated with AVG, were initiated to ripen with exogenous ethylene showed stronger effects than expected. AVG is marketed as ReTain, which contains 15 % w/w AVG, and has been registered for use on apples in several countries. It is applied by spraying directly on fruit in the field (Lurie 2008; Toan et al. 2011). Toan et al. (2010) suggested that 0.8 g L^{-1} of

ReTain sprayed on 'Cavendish' bananas 2 weeks before harvesting, inhibited ethylene biosynthesis and lowered respiratory rate and postharvest losses. The shelf-life of the ReTain treated bananas was extended to 16 days compared to 8 days for the non-treated fruit. Toan et al. (2011) also suggested that the most appropriate of AVG application for bananas was from 74 to 78 days after formation of the last hand of the bunch.

Nitrous Oxide

NO as a key multifunctional-signalling molecule in plants mediating multiple physiological and biochemical responses to biotic and abiotic stresses. It has been used to maintain postharvest quality and extend shelf-life of several fruit and vegetables (Wang et al. 2015a). It also induces plant defence against oxidative stresses by reducing reactive oxygen species accumulation (Wu et al. 2014). Manjunatha et al. (2012) reported that 1 mM sodium nitroprusside, a NO donor, delayed the peak in respiration rate and sugar accumulation and lowered PPO and PAL activity during ripening of 'Cavendish' bananas. Cheng et al. (2009) found that NO reduced ethylene evolution in banana slices during ripening due to the inhibition of ACO activity and the suppression of *MA-ACO1* gene transcription. They also found that NO treatment retained higher contents of acid-soluble pectin and starch and retarded cell wall hydrolases activity, which may account for the reduction of fruit softening. Wang et al. (2015a) found that NO treatment delayed ripening of bananas by 2 days compared to non-treated fruit and preserved chlorophyll content by inhibiting chlorophyll degradation enzymes such as chlorophyllase and Mg-dechelatase activity after cold storage. The tolerance to chilling injury in bananas during cold storage was enhanced by NO treatment through promoting both the enzymatic and non-enzymatic antioxidant systems (Wu et al. 2014; Wang et al. 2015a) and maintaining high levels of energy status and energy metabolism (Wang et al. 2015b). Palomer et al. (2005) stored bananas at 20 °C and showed that ripening was delayed after NO treatment, as judged by ethylene biosynthesis, respiration rate, colour change, acidity and softening. However, they found that these effects were not detectable at 20 % NO concentration, but steadily rose at increasing concentrations above 40 % and appeared to be saturated at 80 %. They concluded that the effects of NO in slowing ripening could be due to its anti-ethylene activity without detrimentally affecting banana quality.

Maleic Acid

Maleic acid (*cis*-butenedioic acid) is a colourless crystalline solid with a double bond in the cis (Z) configuration that has a role as a plant metabolite and can be conjugated to free base compounds or drugs to improve their stability, solubility and dissolution rate. Little work could be found on its effect on fruit ripening, but Copisarow (1935) reported that maleic acid can inhibit ripening in several fruit, including bananas, as well as protecting fruit from decay.

Coatings

Ncama et al. (2018) reported that coating fruit has been successful, not only in reducing water loss and delaying senescence, but also in increasing the antimicrobial properties of the coated produce. Chitosan, gum arabic and several essential oils, have been successfully used as fruit coatings (Thompson et al. 2016). Coatings in the form of natural waxes or synthetic products have been shown to extend the storage life of bananas and plantains by slowing gaseous exchange between the fruit and the atmosphere, and thus delaying the onset of the climacteric. Water loss and respiratory activity were reduced and the duration of the pre-climacteric period of bananas and plantains was increased (Banks 1984; Marchal et al. 1988). Semperfresh and Tal Prolong are names of commercial coatings based on sucrose esters of fatty acids combined with CMC, which has proved effective with bananas. Semperfresh and Tal Prolong are applied by dipping the fruit in a dilute suspension. When allowed to dry they form a very thin, invisible layer which acts to modify the internal atmosphere of the fruit, particularly causing a fall in the O_2 concentration within the fruit, with much less effect on the internal concentration of CO_2. Experiments in which radioactively labelled sucrose ester coatings have been applied to bananas show that these coatings do not migrate into the fruit, but remain on the surface (Bhardwaj et al. 1984). The mode of action of a sucrose-ester coating on the physiology of the ripening banana has been analysed in detail by Banks (1984). In a study of the effect of Semperfresh on the preservation of bananas transported under refrigerated conditions and stored under non-refrigerated conditions (20 °C), Marchal et al. (1988) found ripening to be delayed without any adverse effect on flavour. Similarly, experiments carried out over a period of three years under tropical conditions (Al-Zaemey et al. 1989) showed that coating plantains (*Musa* AAA 'Dominico', 'Harton' and 'Horn') and bananas (*Musa* AAA 'Giant Cavendish') with Semperfresh slowed the major changes associated with ripening (Table 4.3), without affecting the taste of the fruit. A coating similar to CMC was described by Deng et al. (2017). This was cellulose nanofiber-based emulsion (0.3% cellulose nanofiber +1 % oleic acid +1 % sucrose ester fatty acid, w/w on a wet basis), which they found reduced respiration rate and ethylene production in bananas, thus delaying their ripening. The coated bananas ripened normally during 10 days ambient storage; the coating simply delayed the ripening processes. Maqbool et al. (2010) dipped bananas in a composite coating of 10 % gum arabic +1 % chitosan. They showed that after 33 days storage coated fruit had 24 % lower weight loss and 54 % lower TSS than those not coated and were firmer, had higher total carbohydrates and reducing sugars lower than those not coated. The effects of coating bananas with coconut oil (140 or 200 mL) combined with 25 g beeswax and 50 mL sunflower oil during

Table 4.3 Effects of concentration of Semperfresh on plantains that had been ripened to fully yellow at 29 °C. Modified from Al-Zaemey et al. (1989)

	Semperfresh %				
	0	1	1.5	2	LSD $p = 0.05$
Weight loss % day^{-1}	0.62	0.55	0.38	0.26	0.14
Days ripening	3.3	5.3	6.3	6.7	0.36
Peel firmness kg 8 mm^{-1}	6.4	5.6	5.6	4.4	0.35
TSS	18.3	17.5	17.5	18.5	ns

35 days storage compared to non-coated fruit showed better retention of colour and appearance as well as of ascorbic acid (Mladenoska 2013). Truter and Combrink (1990) showed that the storage life of 'Dwarf Cavendish' and 'Williams' was extended when coated with Semperfresh. Banks (1984) reported that coating bananas with TAL Pro-long modified their internal atmospheres by reducing the permeability of the fruit skin to gases. Permeability of control fruit to CO_2 was greater than that of O_2 and ethylene, and this differential permeability was enhanced by coating. This resulted in a depression of the fruit's internal O_2 content which affected ripening without a concomitant increase in the levels of CO_2 which could have proved toxic. The skins of coated fruit lost chlorophyll more slowly than controls and there was a small effect on the accumulation of monosaccharides in the fruit pulp. These effects were associated with depressed rates of respiration and ethylene production in the coated fruit, but the accumulation of acetaldehyde and ethanol was no more rapid than in controls.

Irradiation

Irradiation can have an effect similar to that of coating, but ripening and the organoleptic quality were modified to a certain extent depending on the dose (Thomas 1986). At 13 °C the length of storage was doubled (from 14 to 29 days) by an irradiation of 0.85 kGy, the development of the colour and of softening was blocked, but peel splitting and a loss of aroma are observed (Brodrick and Strydon 1984). PVA is a biodegradable polymer (Matsumura et al. 1999) and tannin it is a renewable natural material (Moric et al. 2014). Bananas were coated with thin films with different ratios of plasticized PVA, CMC and Tannin solutions. Gamma irradiation improved their thermal properties, which provides suitable coating material based on these natural biodegradable polymers for food preservation (Senna et al. 2014). Zaman et al. (2007) treated mature green bananas with 0.30, 0.40 or 0.50 kGy for 5 mins and stored them at room conditions of 25 ± 2 °C and 80 ± 5 % RH. The control bananas ripened within 6 days while the gamma irradiated bananas ripened within 26 days. A minor decrease in the ascorbic acid content was the only adverse effects observed in irradiated bananas and no other major changes occurred in nutritional and organoleptic qualities with no differences detected between the three levels of irradiation.

Extensive work on tropical fruits such as banana, mango and papaya have clearly established that the maturity of the fruits at harvest, the time delay between harvest and irradiation, and the physiological state of the fruit as related to its position in the climacteric at the time of irradiation all can influence the radiation-induced delay of ripening (Thomas 1986). Appreciable delays in ripening and consequent enhancement of shelf-life in bananas and plantains has been reported when pre-climacteric fruits were treated at doses in the range 0.2–0.4 kGy; the extent of delay of ripening is dependent on fruit maturity at harvest and on storage temperature (Maxie et al. 1968, Kao 1971, Thomas et al. 1971, Thomas 1986). Irradiation delayed the onset of the respiratory climacteric and the intensity of the respiratory peak, as evidenced by CO_2 evolution (Maxie et al. 1968, Thomas et al. 1971). Maximum delay of ripening has been observed with fruits of lower maturity (threequarters to full three-quarters); as fruit maturity increased, a progressive decrease in ripening delay and shelf-life occurred (Kahan et al. 1968, Thomas et al. 1971). The optimal radiation dose for inhibition of ripening, and the maximum dose the fruits can tolerate without exhibiting phytotoxicity or radiation-induced injury, seem to differ among cultivars and even for the same cultivar grown in different geographic areas. However, irrespective of cultivar differences, doses exceeding 0.5 kGy invariably resulted in browning or blackening of the skin in pre-climacteric bananas, while doses of 1 kGy and more often caused splitting and softening of fruits. Fruits that are already in the climacteric state can tolerate doses as high as 2 kGy without any external manifestation of radiation injury. The browning of skin occurring in pre-climacteric fruits exposed to higher doses has been attributed to increased PPO activity (Thomas 1986; Surendranathan and Nair 1972).

Following the observation that irradiated bananas, when exposed to ethylene, take longer periods to reach a comparable stage of ripeness than do non-irradiated fruits, it has been postulated that the inhibition of ripening caused by irradiation involves a decreased sensitivity to the ripening action of ethylene (Maxie et al. 1968, Thomas et al. 1971). Gamma radiation. Alterations in carbohydrate metabolism by regulating certain key enzymes of the glycolytic pathway, TCA cycle and gluconeogenic pathway, and the resultant interference with ATP production required for various synthetic processes during ripening have been reported as some of the main causes for the delay of ripening in irradiated bananas (Surendranathan and Nair 1973 and 1980). Predominance of the pentose phosphate pathway, accompanied by a gradual activation of fructose-1-6- diphosphatase, decreased succinic dehydrogenase activity and the operation of glyoxalate shunt pathway, as evidenced by increases in isocitrate lyase and malate synthetase, have been reported in 'Dwarf Cavendish' bananas irradiated in the pre-climacteric stage (Surendranathan and Nair 1972, 1976). An impairment of succinic dehydrogenase activity has been observed in mangoes following irradiation (Thomas et al. 1971). In mature mango fruits of the cultivar 'Haden', the increase in NADP-malic enzymatic activity, usually observed during ripening, was significantly diminished but not delayed by irradiation at 0.75 kGy (Dubery et al. 1984). An exponential decrease in the activity and significant differences in some of the allosteric properties and kinetic parameters Vmax and Km of NADP malic enzyme purified from mango fruit have been reported when irradiated *in vitro*

(Dubery et al. 1984). However, it is unlikely that these effects observed *in vitro* are the same as in intact fruits irradiated with the same doses.

An increase in ascorbic acid level was observed in 'Gros Michel' bananas irradiated at 0.25–0.5 kGy, which was attributed to increased extractability (Maxie et al. 1968), whereas in five banana cultivars grown in India, irradiation at levels optimal for delaying ripening (0.3 kGy) did not significantly affect ascorbic acid levels (Thomas et al. 1971). In 'Alphonso' mangoes grown in India, the temperature at which fruits were stored and ripened was found to influence the changes in ascorbic acid levels during ripening. Increased niacin levels were observed in 'Gros Michel' bananas irradiated at 0.3 and 0.5 kGy, whereas the thiamin content was unaltered at these doses (Maxie and Sommer 1968).

Temperature

In 1928 the Low Temperature Research Station was established in St Augustine in Trinidad at the Imperial College of Tropical Agriculture. The initial work was confined to "...... improving storage technique as applied to Gros Michel [bananas]...........for investigating the storage behaviour of other varieties and hybrids which might be used as substitutes for Gros Michel in the event of that variety being eliminated by the epidemic spread on Panama Disease" (Wardlaw and Leonard, 1940).

Symptoms of chilling damage to bananas include a reddish-brown streaking in the skin associated with the vascular bundles, reduced latex flow, slow colour development during ripening, a dull or greyish yellow colour in ripe fruit and brown skin in advanced stages when they were exposed to 4–6 °C. More subtle changes are the reduced production of volatiles in the ripening fruit and disturbances to the evolution of CO_2 and ethylene. A delay in the climacteric rise in CO_2 occurs and multiple peaks appear instead of one (Murata 2006). Beyond 25 °C the duration of the pre-climacteric is greatly shortened, and fruit quality is altered, because of modifications to fruit metabolism during ripening. Above 35 °C the development of the peel and of the pulp is desynchronized, with softening of the pulp proceeding faster than the colouring of the peel. This results in fruit with a soft pulp but a green peel. The fruits are then in a condition known as 'cooked/boiled green'. Above 48 °C the climacteric is not triggered and ripening is effectively blocked. If the temperature is too low it can cause chilling injury even at 13 °C (Fig. 4.4). "Under peel discoloration" is a mild form of chilling injury that occurs to bananas when they are stored at just below the critical temperature/time combination. Longer exposures and lower temperatures can result in more severe injury symptoms, as described above, including failure to ripen, dull or greyish to the peel of ripe fruit and in severe cases the peel turns dark brown or black, and even the flesh can turn brown and develop an off taste.

Harvey (1928) reported that "relatively high temperatures speeds up ripening when they are allowed to ripen on the plant". He reported that for green 'Gros Michel', 4 or 5 days may be required for ripening at 65–70 °F, but at 80 °F this time

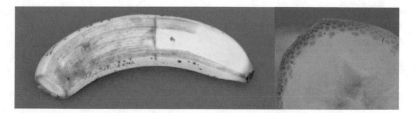

Fig. 4.4 Chilling injury in 'Cavendish' in the form of under peel discoloration (UPD). (Photographs Allen Hilton, Cranfield University)

may be decreased, but at 60°F. ripening will be much slower. Generally, it has been found that ripening will be better throughout the pulp if the temperature is not much above 65 °F. At that time Harvey (1928) reported that "high temperature has a tendency to shrivel the fruit. However. jobbers frequently use high temperatures when fruit is in demand because the fruit reaches a marketable color much more quickly than at 65 °F. High temperature with attendant high humidity favors the development of molds and the blackening of fruits, and hastens the rotting of the fruit stalk. A blackened skin gives a bunch of bananas a poor appearance. The rotting of stalks becomes a serious matter when the decay of the fingers allows the fruits to break loose from the hand. Reynolds suggests the following temperatures for the control of ripening rooms: 72 °F. (or over) Danger of cooking 68 °F. Fast or forced ripening 66–62 ° F. Normal ripening 60 °F. Slow ripening 58 °F. Holding green bananas 56 °F. Holding ripe bananas. There is some danger of chilling even ripe fruits at temperatures below 54 °F. Green bananas may show effects of chilling at slightly higher temperatures. Incidentally, the banana will not freeze until the temperature gets to a few degrees below the freezing point, 32 °F. However, badly chilled fruits that have not been frozen are liable to turn black. If only slightly chilled they may show a gray-green appearance after ripening. In ripening tomatoes, the use of high temperatures (above 90 °F.) is to be avoided as far as possible, because the fruit rots badly. Usually there are enough fungus spores on the fruits to cause rotting if conditions are favourable. It is a problem, then, with the use of heat alone, to avoid rotting discoloration, shrivelling, and poor flavor at high temperature, or to avoid the long time required at low temperature. The time required to ripen 'Cavendish' bananas by heat alone is so great as to exclude this genotype from commercial use."

Since bananas suffer from chilling injury and recommended storage temperatures are essential the minimum temperature that can be used before chilling injury occurs, temperatures close to that minimum would be the best for maximizing their postharvest life. The skins of fruit stored at below about 13 °C gradually darkened with increasing relative electrolyte leakage. Under peel discolouration is a milder form of chilling injury and can also occur during growth when night temperatures drop to below 13 °C and is caused by latex coagulation in the peel laticifers and subsequent browning of the latex by phenolic oxidation (Fig. 4.4). Chilling injury of bananas stored below 13 °C was also reported to be associated with a decrease in ethylene binding. The ability of tissue to respond to ethylene is evidently reduced, thereby resulting in failure to initiate ripening (Yueming Jiang et al. 2004). However, chilling injury is time as well as temperature dependent that is why some authorities give maximum time or a journey time. Recommendations for optimum storage of green bananas include the following:

'Giant Cavendish'

>   12.8–14.4 °C with 85–90 % RH (Pantastico 1975)
>   13–14 °C (Thompson and Burden 1995)
>   14 °C for 'Grand Naine' (Robinson and Saúco 2010)

'Dwarf Cavendish'

>   11.5–11.7 °C (Wardlaw 1937)

'Gros Michel'

>   11.5–11.7 °C (Wardlaw 1937)
>   11.7 °C (Hardenburg et al. 1990)

'Lacatan'

>   14–14.4 °C (Wardlaw 1937)

'Latundan'

>   14.4–15.6 with 85–90 % RH for 3–4 weeks (Pantastico 1975)

Non-specified varieties

>   13 °C (Purseglove 1975)
>   12.8–15.6 °C with 85–90 % RH for 4 weeks (Pantastico 1975)
>   12–14 °C and 90–95 % RH for 2–3 weeks (Mercantilia 1989)
>   13 °C and 85–90 % RH for 10–20 days (Mercantilia 1989)
>   13 °C and 85–90 % RH for 10–20 days (Snowdon 1990)
>   13–15 °C and 85–90 % RH for 1 month (Snowdon 1990)
>   13.3–14.4 °C and 90–95 % RH for all cultivars except Gros Michel (Hardenburg
>   et al. 1990)
>   14.4 °C and 85–95 % RH for 7–28 days (SeaLand 1991)
>   12–14 °C and 85–95 % RH for 15 days (Tchango et al. 1999)

Humidity

Water deficit in plant tissues may stimulate ethylene production and as a consequence there can be an increase of tissue respiration rate (Yang and Pratt 1978). Exposing fruit postharvest to low humidity may also result in sufficient stress to the fruit to initiate ripening, as was shown by for plantains (Table 4.4).

Storage humidity was shown by Wardlaw and Leonard (1940) to affect respiration rate of 'Gros Michel' bananas in that at 29.4 °C the climacteric was earlier for those in 100 % RH compared to those in 70 % RH (Fig. 4.5). In experiments it was

Table 4.4 The effects of the humidity of the incoming air in a flow through system on the ripening of freshly harvested plantains in ambient conditions (23–33 °C) in Jamaica. Penetrometer force was the force required to inject a 5 mm Magness and Taylor pressure tester probe (adapted from Thompson et al. 1972)

Storage humidity	Colour score	Weight loss %	Total soluble solids	Pulp:peel ratio	Penetrometer force	
					Skin	Pulp
100 % RH	2.3	0.9	10.6	1.9	12.0	6.6
0 % RH	7.0	18.4	24.9	3.5	8.2	1.8

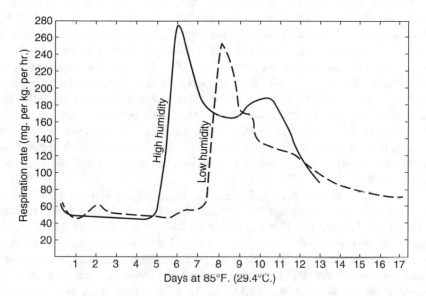

Fig. 4.5 Effects of humidity on respiration rate of 'Gros Michel'. Source: Wardlaw and Leonard 1940

shown that plantains stored in cartons without a packing material lost weight rapidly during ripening and became blackened and shrivelled. Those packed in dry coir dust lost less weight, and those in moist coir dust gained weight; no shrivelling was seen following the latter two treatments and there was less skin blackening (Table 4.2). The fruits in coir dust remained green for long periods before ripening rapidly (Thompson et al. 1974). However, they took up moisture and would eventually split if stored for too long.

Littmann (1972) found that water stress decreased the pre-climacteric life of 'Giant Cavendish' bananas and showed a depression in the climacteric peak of respiration rate. Finger et al. (1995) reported that fresh mass loss of 5 % and higher promoted shortening of the pre-climacteric life of bananas and induced a decrease of maximum rates of respiration and ethylene production during climacteric ripening. Pre-climacteric ethylene production was stimulated by water stress. Fruit with a mass loss of 20 % ripened abnormally with a decreased pulp softening and excessive brown colour of skin. Akkaravessapong et al. (1992) examined the effect of approximately 50 % RH, approximately 70 % RH and approximately 90 % RH on the susceptibility of bananas to mechanical damage from 2 days after harvesting until 5 days after ripening had been initiated. Humidity did not influence susceptibility to mechanical injury, the tissues damaged at 50 % RH dried to a black colour while those damaged at 90 % RH remained light brown. George and Marriott (1985) found that in plantains an increase of weight loss of 0.5 % per day resulting from low storage humidity was typically accompanied by a decrease of 65 % in the potential storage life. They also found that treatment with gibberellin extended the pre-climacteric period when subsequent storage was under high humidity (unstressed conditions), but had no effect under low humidity (water stressed conditions). Ullah et al. (2006) observed that high humidity delayed the ripening of bananas and consequently increased their shelf-life. Storage of bananas submerged in water delayed the climacteric phase of bananas in a similar way to controlled atmosphere storage but they did not complete their ripening processes properly. Storage of bananas submerged in water resulted in splitting of bananas due to excessive intake of water. Microbial spoilage of water-stored bananas emerged as serious problem against successful application of this technology. Reduced dissolution of O_2 into water affected the ripening process, and in order to attain the best quality and increased shelf-life, water storage was found to be not suitable for bananas but storage in 100 % RH showed better results.

The effects of storage in polyethylene film bags on the postharvest life of fruits may also be related to moisture conservation around the fruit as well as the change in the CO_2 and O_2 content. This was shown by Thompson et al. (1972) for plantains where fruit stored in moist coir or perforated polyethylene film bags had a longer storage life than fruits that had been stored unwrapped. But there was an added effect when fruits were stored in non-perforated bags, which was presumably due to the effects of the changes in the CO_2 and O_2 levels (Table 4.2). So, the positive effects of storage of fresh pre-climacteric fruits in sealed plastic films may be, in certain cases, a combination of its effects on the CO_2 and O_2 content within the fruit and the maintenance of high moisture content. The effect of moisture content is more likely to be a reduction in stress of the fruit that may be caused by a rapid rate of water loss in non-wrapped fruit. This in turn may result in increased ethylene production to internal threshold levels that can initiate ripening.

George and Marriott (1982) reported that the lower the humidity, the greater the loss of water and the shorter is the duration of the pre-climacteric period, with plantains being more sensitive than other *Musa* clones tested. Low humidity brought forward the production of ethylene and the climacteric rise in respiration, but its effect on the level of ethylene production varied with genotype. For example, in plantains (*Musa* AAB 'Apem') the production of ethylene and the level of ACC were increased (George and Marriott 1985), while Nair and Tung (1988) observed the opposite for 'Pisang Mas' (*Musa* AA). For all the clones studied, a reduction in humidity did not alter the rate of respiration, but affected the pulp and peel ratio, peel colour, pulp softening and in TSS (Broughton et al. 1978; George and Marriott 1985; Nair and Tung 1988). These effects were shown for plantains (Table 4.2).

Chapter 5
Initiation of Ripening

Burg and Burg (1967) reported that carbon monoxide, butylene, propylene and acetylene all had biological activity similar to that of ethylene when tested with the pea straight growth test. They also stated that a pre-requisite for effectiveness in initiating banana ripening appears to be that they must be unsaturated, that is, have a double or triple bond between carbon atoms. Dhembare (2013) also named several chemicals that had been used in banana ripening including calcium carbide, acetylene gas, carbon monoxide, potassium sulfate, Ethrel, potassium dihydrogen arthoposphate, putrisein, oxytocin and photoporphyrinogen. However, gases other than ethylene that have been shown to initiate ripening of bananas were all considerably less effective than ethylene. Kader A.A (1992 quoted by Siddiqui and Dhua 2010) reported the comparative effectiveness of ethylene and other compounds in initiating fruit ripening where ethylene = 1, propylene = 130, vinyl chloride = 2370, carbon monoxide = 2900, acetylene = 12,500 and 1-butene = 140,000, although it is difficult to reconcile some of these figures with published experimental work.

Ripening of individual bananas may vary within a bunch and even within a hand (Fig. 5.1), therefore for most commercial practice fruit are harvested at the preclimacteric stage and initiated to ripen by application of a chemical exogenously. In some local markets it is actually preferred that all the bananas in one hand do not ripen at the same time, because this allows them to be eaten in sequence since many people like to eat only one or two bananas each day so initiating them to ripen all at the same time is a disadvantage.

In the experience of the authors many people believe that fruits that are not initiated to ripen exogenously by a chemical taste better. This has been studied and results have proved inconclusive. Limited tests carried out with academic Food Technologists in Thailand also proved inconclusive, but generally slightly favoured those that had been initiated to ripen exogenously (Jiraporn and Thompson 2019 unpublished). Sonmezdag et al. (2014), based on sensory analysis, reported that banana not exposed to exogenous ethylene were preferred because of their better fruity aroma and general impression attributes.

© The Author(s), under exclusive license to Springer Nature Switzerland AG 2019
A. K. Thompson et al., *Banana Ripening*, SpringerBriefs in Food, Health,
and Nutrition, https://doi.org/10.1007/978-3-030-27739-0_5

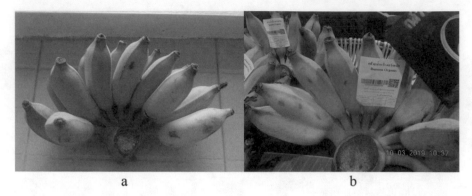

a   b

Fig. 5.1 'Namwa' (*Musa* AAB) ripened naturally on sale at the road side (**a**) and in a supermarket (**b**) in Bangkok in March 2019.

A comparison of the quality of bananas ripened with or without exogenous ethylene to the same skin colour was made by a descriptive analysis panel (Scriven et al. 1989). They found the pulp of the fruit ripened without exogenous ethylene was less green, more fruity and softer than the exogenous ethylene ripened fruit. The fingers of the same hand of a bunch may also initiate to ripen at different times. Also, the production of endogenous ethylene by fruits placed in the same conditions varies between genotype. For example, Karikari et al. (1979) showed that under conditions where 'Cavendish' (AAA) produced 2.1–2.5 μL^{-1} kg hr.$^{-1}$ ethylene, tetraploids (*Musa* AAAA) produced 6.7 μL^{-1} kg hr.$^{-1}$ ethylene and 'Apem' (AAB) produced 34.8 μL^{-1} kg hr.$^{-1}$ ethylene. When a local Thai farmer, who grows bananas on a small scale, was asked which method he uses for initiating ripening he replied that "there are 2 reasons for non-using ethephon to activate banana ripening. The first, using ethephon makes more rapid spoilage on banana than calcium carbide. He said, it may come from water for preparation ehtephon solution. Secondly, using ethephon takes shorter time than calcium carbide for banana ripening. It is too fast for sale. When he would like to sale banana, he needs time for waiting customers. If the banana goes ripening too fast, it will go to spoilage too fast as well. This situation is not good for him (sale person)" (translated by Jiraporn Sirison). Jiraporn and Thompson (unpublished, 2019) ripened 'Namwa' bananas at 29–31°C and 71–86% RH either with ethylene or left them to ripen "naturally". Those initiated with ethylene were fully ripe and an even yellow colour after 4 days while those ripened "naturally" took 8 days to ripen and still had green tips on each finger, which persisted even when they were over-ripe and very soft after 11 days.

Ethylene

Ethylene is used in fruit ripening rooms, which is a colourless gas with a sweetish odour and taste, has asphyxiate and anaesthetic properties and is flammable. Its flammable limits in air are 3.1–32% volume for volume and its autoignition temperature is 543 °C. Care must be taken when the gas is used for fruit ripening to ensure that levels in the atmosphere to not reach 3.1%. As added precautions all electrical fitting must be of a "spark-free" type and warning notices relating to smoking and fire hazards must be displayed around the rooms.

Ethylene was first identified by Neljubow (1901) and has been known to have physiological effects on crops for over 100 years. F.E. Denny used ethylene for ripening bananas, tomatoes and pears and subsequently took out a patent (Denny 1923). Ethylene was first identified as a volatile chemical produced by ripening apples by Gane (1934). It was assumed that ethylene was produced only in climacteric fruits during the ripening phase, but with the development of chromatographic analytical techniques, it is clear that all crops are able to produce ethylene under certain conditions (Hulme 1971). In his review, Curry (1998) reported that ethylene not only initiates ripening of climacteric fruit but can also induce fruit abscission, induce flowering, promote seed germination, break dormancy, promote root initiation, delay sprouting and promote plant dwarfing.

Bananas are sensitive to physiological levels of ethylene as low as 0.3–0.5 µL L^{-1} if the O_2 and CO_2 levels are similar to those found in outside fresh air (Peacock 1972). Biale (1960) and Gane (1936) reported that 0.1–1 µL L^{-1} levels of ethylene were required to initiate ripening. The main factors affecting response to exogenous ethylene are: fruit maturity, time after harvest when ethylene exposure began, temperature and the length of exposure to ethylene. 'Cavendish' bananas have been shown to be more sensitive to ethylene at 35 °C than at 20 °C (Seymour et al. 1985). Traditionally in commercial rooms the fruits were maintained in an atmosphere of 1000 µL L^{-1} ethylene for 24 h, generally between 16 and 18 °C and humidity as high as possible (around 95% RH) (Thompson and Seymour 1982; Nelson nd). As was described earlier, the various changes that occur during ripening might not all occur simultaneously. For example, in pears a differential effect on the initiation of individual ripening processes has been found. Pears required 0.08 µL L^{-1} ethylene to initiate softening while initiation of the respiratory climacteric required 0.46 µL L^{-1} (Wang and Mellenthin 1972).

The effect of endogenous ethylene on ripening is universally accepted to follow a time–concentration relationship with the higher the concentration of ethylene and the longer the exposure time to ethylene, the faster the initiation of ripening (Wills et al. 1998). Recommended concentrations for the induction of ripening of avocados, bananas, honeydew melons, kiwifruit, mangoes, stone fruits and avocados are between 10 and 100 µL L^{-1} (Saltveit 1999). Knee (1985) reported a threshold level of 0.1–0.5 µL L^{-1} to initiate the ripening of avocado, banana, honey dew melon and pear. However, studies on kiwifruit have found that ripening was initiated by levels of 0.01 µL L^{-1} (Mitchell 1990). Peacock (1972) speculated that for bananas there is no effective threshold level of ethylene that induces ripening, although he examined short term exposure to ethylene levels >0.1 µL L^{-1} rather than longer exposure to concentrations <0.1 µL L^{-1}. 'Cavendish' and 'Lady Finger' bananas were stored at 15, 20 and 25 °C in an atmosphere containing 0.001, 0.01, 0.1 and 1.0 µL L^{-1} ethylene in air and as expected, green life increased as the temperature and ethylene concentration decreased (Wills et al. 2014). Postharvest exposure to ethylene can have negative effects on crops including: apples – scald, aubergines - brown spots, potatoes – sprouting, grapes – mould, onions and garlic – odour, broccoli – yellowing, carrots – bitterness and leafy vegetables - loss of green colour.

Cylinders

Large gas cylinders containing ethylene under pressure can be used in rooms to initiate banana ripening (Fig. 5.2). One method of application is to calculate the volume of the ripening room and from a cylinder of ethylene gas, enrich the ethylene level until it reaches the required amount. This is done with a stop watch and flow meter attached to the pipe connecting the room with the gas cylinder (Fig. 5.3). The problems associated with this method are that ethylene can be explosive in air and, although only 0.1% is commonly used a mistake can easily be made and ethylene concentrations may reach explosive levels. The second problem is that if the ripening room is not completely gas tight then the ethylene may leak and the level fall below the threshold that is required to initiate ripening. The former problem is

Fig. 5.2 Large cylinders of ethylene outside a banana ripening room in Brazil

Fig. 5.3 Initiation of
ripening using an ethylene
cylinder

partially addressed by diluting the ethylene in the cylinders with nitrogen giving
about a 5% ethylene concentration which gives a potential greater margin of error if
a mistake is made when the gas is being applied. Also, the ethylene gas may be
applied in small cylinders which are called 'lecture tubes'. These contain a small
amount of pure ethylene (often 35 L) under pressure. The whole of the contents of
the tube is thus released into the room. If, for example, the ripening room had a
volume of 70 m^3 and a desired ethylene concentration of 1000 μL L^{-1} was required
then two 35 L lecture tubes would be discharged into the room.

Ripening of bananas in Australia was reported to be at 13–14 °C for 2–7 days,
with the length of time depending on the distance of the wholesaler from the port of
entry and on market demand. To ripen the bananas, the polythene bag in which the
fruit was stored, is opened and ethylene gas applied to the fruit. Controlled ripening
is undertaken at 14.5–21 °C. Humidity is generally maintained at 85–95% RH in the
early stages of ripening and, once a trace of colour appears, it is reduced to 70–80%.
Ripening rooms are usually gassed with ethylene on each of two successive days
(Story and Simons 1999).

Catalytic Generators

Ethylene generating equipment, which gives slow release of ethylene over several hours, is now available; it addresses both the problems of room leakage and minimizes the hazards of use. Suitable equipment is manufactured commercially which gives a constant stream of ethylene over a 16 h period (Fig. 5.4). The number of ethylene generators required depends on the size of the room. Catalytic generators probably generating ethylene by heating ethanol in a controlled way in the presence of a copper catalyst. Care should be exercised in doing this because of the inflammability of the alcohol. They are patented and careful controls need to be established to ensure when such a process is set up it is completely safe.

Ethrel

Compounds, in which the fruit can be dipped and decompose in or on the fruit to release ethylene can have the advantage of easy application. Etacelasil (2-chloroethyl-tris-[ethoxymethoxy] silane) or ACC (1-aminocyclopropane-1-carboxylic acid) have not been used practically for the application of ethylene but 2- chloroethyl phosphonic acid is marketed as Ethrel and also called Ethephon. Reid (2002) reported that the use of Ethrel in the USA is approved on some crops in certain States and circumstances including postharvest fruit ripening of bananas, tomatoes and peppers. Ethrel reacts with water releasing ethylene but it is also hydrolysed in plant tissue to produce ethylene, phosphate and chloride.

$$Cl-CH_2-CH_2-PHO_3^- \xrightarrow{OH^-} CH_2=CH_2+PH_2O_4^-+Cl^-$$

Fig. 5.4 Ethylene generator being used in a ripening room. Reproduced with permission of Greg Akins, Catalytic Generators Inc., Norfolk, VA. USA

In practice ethylene can also be released from Ethrel by mixing it with a base such as sodium hydroxide. For example, Ethrel C (a proprietary brand) will release 93 g from 1 L or 74.4 L of ethylene gas per litre of Ethrel. It has been used in this way to initiate fruit to ripen by placing containers of Ethrel in a gas tight room containing the fruit and then adding the base to the containers. This method was used in commercial ripening rooms in the Yemen, using mainly 'Dwarf Cavendish' and some 'Poyo'. Buckets of dilute sodium hydroxide solution were placed throughout the ripening rooms. When all these were in place measured amounts of Ethrel are added to each bucket, which gave an instant release of ethylene gas into the room. The mixture was 80 mL Ethrel and 1 L of sodium hydroxide for every 300 m^3 of room, in theory the temperature of the pulp of the banana should be 20–22 °C, but in practice the pulp temperature was 29–30 °C. The rooms remained closed for 3 days, but rubber gaskets on the doors were pulled away in places and had been damaged by fork lift trucks so they were far from gas tight. On removal after 3 days the fruit were quite soft, but still green and lacking in flavour. A comparison was made between using this method and initiating ripening with calcium carbide. The Ethrel method was found to be some 50 times more expensive than using calcium carbide but was more effective than calcium carbide (Thompson 1985).

Venkata Subbaiah et al. (2013) ripened bunches of 'Grand Naine', half the fruit were dipped in Ethrel and the other half not dipped, at room temperature (25–29 °C and 79–89% RH). The Ethrel treated fruit ripened in 4 days with excellent quality attributes, whereas non-treated fruit took 10 days to ripen. Ram et al. (2009) dipped unspecified genotypes of bananas in various concentrations of Ethrel (1000, 2000, 3000 or 4000 ppm) and found little difference between the concentrations, and fruit dipped in 1000 ppm for 5 minutes started to soften in 3 days (compared to 9 days for the control) and became ready to consume in 5 days with shelf-life of 8 days. They were considered to have an excellent flavour. In Thailand Ethrel is used for fruit ripening as well as calcium carbide. Ethrel was purchased locally (March 2019) for THB 350 per litre. Two overflowing capfuls were added to half a bucket of water (Fig. 5.5) and the hands of bananas were dipped in the bucket so that they were fully immersed. The amounts used by local traders were measured, and it was found that they added about 30 mL of Ethrel to 7.5 L of water which gave an approximately 250 ppm solution. The recommendation on the bottle of Ethrel for ripening bananas was 500–1500 ppm, so they were using considerably less than the recommended level. After dipping the fruit were placed on a table to ripen at a constant temperature of 29–31 °C (Fig. 5.6). Colour changes were perceived in the Ethrel treated hands after 2 days. After 4 days the non-treated hands appeared to be exactly as they were 4 days earlier with latex exuding from the skin when they were peeled and being hard, difficult to peel and tasting bitter. The Ethrel treated hands at the same time were of excellent texture, flavour and easy to peel (Fig. 5.7). After the non-treated fruit were tested 11 days after harvesting, they were overripe and very soft with the fingers falling away from the crown, but they were still good to eat. However, there was still a very small amount of green at the distal end of the peel of fingers. This is in contrast with the fruit initiated to ripen with Ethrel that had not green in the peel when they were fully ripe and soft (after 4 days) (Fig. 5.8).

Fig. 5.5 Preparing Ethrel for dipping bananas in Thailand

Fig. 5.6 'Kluai Hom Thong' *Musa* AAA were used in this test at 5.00 pm on 21 March 2019 on left is a banana hand 1 h after being dipped and on the right a hand that had not been dipped

Fig. 5.7 'Kluai Hom Thong' *Musa* AAA 25 March 2019 8.00 am 4 days after Ethrel treatment. The one on the left appeared to be exactly as it was 4 days ago with latex exuding from the skin when it was peeled and being hard, difficult to peel and tasting bitter. The one on the right was of excellent texture, flavour and easy to peel

Fig. 5.8 'Kluai Hom Thong' *Musa* AAA harvested 21 March. Not treated to initiate ripening and stored at about 27–31 °C and 71–86% RH and photographed on 28 March 2019 8.00 am 7 days after harvest, 29 March 2019 8 days after harvest and 30 March 2019 9 days after harvest, 31 March 2019 10 days after harvest, 1 April 2019 11 days after harvest

Ethrel can contain impurities including 1,2-ethanediylbis (phosphonic acid) and monochloroethyl ester of (2-chloroethyl)-phosphonic acid (Segall et al. 1991, EPA 1988). The latter may degrade to monochloroacetic acid (EPA 1988). Islam et al. (2016) reported that both Ethrel and ethylene glycol samples they tested contained trace amounts of sulphur compounds. Ethrel has several many uses and has been used commercially to initiate flowering in pineapples, anti-lodging in cereals, stimulation of latex flow in rubber, enhanced degreening of citrus fruits as well as initiation of ripening in climacteric fruit.

Encapsulation

Ho et al. (2011) developed molecular encapsulation of ethylene into α-cyclodextrin and tested ethylene release from the capsules at 52.9, 75.5 or 93.6% RH and 45, 65, 85 or 105 °C. Kinetics analysis, based on Avrami's equation (describes the kinetics of crystallisation), showed that the release of ethylene from the capsules increased as humidity and temperature were increased. They subsequently investigated using deliquescent calcium chloride and magnesium chloride to release ethylene from an ethylene-α-cyclodextrin inclusion complexes (IC) powder at humidities between 11.2 and 93.6% RH at 18 °C (Ho et al. 2015). They found that when the IC powder and deliquescent salts were mixed at a ratio of 1:5, respectively, the $CaCl_2$ and $MgCl_2$ started to deliquesce at 32.7% RH when the IC powder dissolved in the concentrated salt solutions to release ethylene. Increasing the humidity accelerated the release rate. Maximum release of ethylene gas was achieved after 24 h at 75.5 and 93.6% RH for both IC powder-deliquescent salt mixtures. The deliquescent salts proved to be a simple option for releasing ethylene gas from the IC powder and is being marketed as Ripestuff™ (University of Queensland 2017). The powder is an ethylene-α-cyclodextrin inclusion complex.

When tested on mangoes (Ho et al. 2016) in the laboratory, for in-transit ripening over two seasons they found that ethylene gas started to be released from the IC powder in 2 h and complete release was achieved in 24 h. Assessments of fruit colour and firmness showed that encapsulated ethylene and commercial grade ethylene from pressurised cylinder similarly shortened the ripening time to 9–10 days (after harvest) for treated fruit as compared with 15 days for untreated mangoes.

Acetylene

Acetylene gas can be used in place of ethylene and has the advantage of being inexpensive and readily available in many developing countries. As reported above, ethylene generated from Ethrel can cost 50 times more than acetylene generated from calcium carbide (Thompson 1985).

Thompson and Seymour (1982), Smith et al. (1986) Smith and Thompson (1987), tested the effects of acetylene on initiation of ripening of pre-climacteric 'Giant Cavendish' bananas. They used pure acetylene, not calcium carbide, and obtained the required concentrations of acetylene from cylinders of high purity mixtures of acetylene in air, specially prepared by British Oxygen Company, at various concentrations. Acetylene was shown to be effective in initiating ripening of bananas. Ethylene initiated ripening at 19 °C after exposure for 24 h at 0.01 mL L^{-1}, but 1 mL L^{-1} of acetylene was required to give the same effect at that temperature (Thompson and Seymour 1982). Smith et al. (1986) showed that the effects of acetylene on bananas was proportional to temperature and exposure time (Fig. 5.9) where exposure of fruit to 1 mL L^{-1} of acetylene for 4 h did not initiate ripening at between 20 and 30 °C but it did initiate ripening at 35 °C.

With 8 h exposure to acetylene, fruit were not initiated at 20 °C, were partially initiated at 25 °C and were fully initiated at 30 °C. Smith et al. (1986) observed an interaction between the pulp temperature of bananas and their response to acetylene in that, over a temperature range of 14–35 °C, the higher the temperature the shorter the exposure time, or the lower the concentration of acetylene was required. The period of exposure to acetylene was also shown to affect initiation of ripening, with longer exposures, up to 24 h, requiring a lower concentration of acetylene or a lower

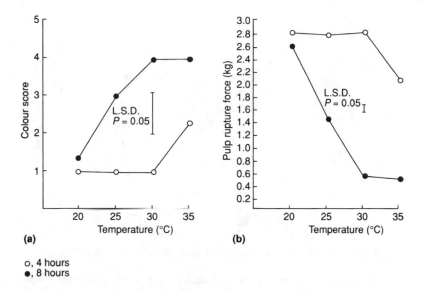

o, 4 hours
●, 8 hours

Fig. 5.9 Effects of exposure for either 4 or 8 h to 1 L per cubic metre of acetylene at various temperatures on the ripeness of bananas after 6 days at 20 °C. Where a = peel colour score; b = pulp firmness. Smith et al. (1986)

temperature to initiate ripening (Fig. 5.10). Bananas exposed to acetylene at the lower temperatures (14–16 °C) were shown to ripen normally with respect to pulp TSS and pulp texture (rupture force), although the co-ordination of the degreening response was lost in some fruit exposed to acetylene for 72 h. They found, however, that the temperature at which bananas were initiated to ripen did not affect their flavour.

Burg and Burg (1967) reported that acetylene had a lower biological activity than ethylene, when tested with the pea straight growth test. However, they found that in order to initiate ripening of bananas acetylene should be applied at 2.8 mL L⁻¹. In mangoes it was found that at 25 °C and with a 24 h exposure time, at least 1 mL L⁻¹ of acetylene or 0.01 mL L⁻¹ of ethylene was required to initiate ripening (Medlicott et al. 1987). They also found that treatment with 0.01 mL L⁻¹ acetylene resulted in limited softening of mangoes but had no effect on the other ripening changes analysed.

Calcium Carbide

Calcium carbide reacts with water to produce acetylene:

$$CaC_2 + 2H_2O \rightarrow Ca(OH)_2 + C_2H_2$$

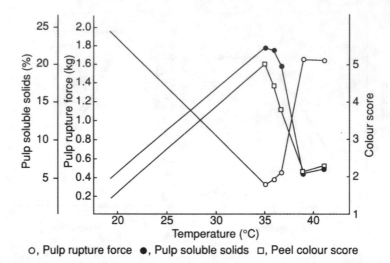

o, Pulp rupture force ●, Pulp soluble solids □, Peel colour score

Fig. 5.10 The effects of exposure at 20 °C for 6 days to 1 mL L⁻¹ of acetylene at different temperatures on the ripening of bananas. Source Smith et al. 1986

In many third world countries CaC_2 is used as described above. Acetylene is used commercially to initiate ripening in many less developed countries in the form of CaC_2 because it is cheaper than ethylene sources and easier to apply in ripening rooms. Calcium carbide is a by-product of the iron and steel industry and the material available contains impurities. Technical grade calcium carbide of regular chip size of 4–7 mm conforming to British standard BS 642 (1965) is the type most commonly used. If it were pure calcium carbide then 1 kg would produce 300 L of acetylene gas. The gas is released when the calcium carbide is exposed to moisture. The reaction can be violent so the way it is commonly applied is to wrap small amounts (just a few grams) in twists of newspaper and put these among the bananas to be ripened (Fig. 5.11). The high humidity reacts with the calcium carbide giving a slow release of acetylene. Where large quantities of acetylene are required quickly the small amounts of calcium carbide can be dropped carefully into large buckets of water. Morton (1987) reported that in 1874 in the Azores it was accidentally discovered that smoke would bring pineapple plants into bloom in 6 weeks and in 1936 acetylene were employed to expedite uniform blooming. Some growers deposited calcium carbide in the crown of each plant that released acetylene when it got wet.

There is some indication that different varieties of banana require different amounts of calcium carbide to initiate ripening. Acedo and Bautista (1993) showed that 'Latundan' (*Musa* ABB) and 'Saba' (*Musa* BBB) bananas exposure of 5 g of calcium carbide for 1 day "effectively enhanced ripening". 'Latundan' harvested at full maturity required a lower level of calcium carbide then 'Saba' harvested at the full ¾ stage.

In Thailand farmers buy a small amount of calcium carbide in a local shop (perhaps 100–150 g) that has been broken down into small packages by the shop owner.

Fig. 5.11 Calcium carbide (27 g) wrapped in paper and inserted in a box of 'Namwa' bananas (4.2 kg) to initiate them to ripen in Thailand

Farmers are charged 10 TBT for a package, but it is possible to buy a kg sealed in a plastic bag for 60 TBT. The farmer then puts a small amount (20–30 g) in a paper with about 60 banana fingers. This ratio will vary depending on the size of the fingers and whether the carton provides an effective barrier to loss of acetylene, that is, whether it is well padded with paper and cardboard. The judgement of all these factors is based on individual farmers' experience. It is impossible to calculate the dose of acetylene received by individual fingers, but from observations of bananas that have been initiated to ripen in this way, it works,

Sensory Analysis

Results reported by Nura et al. (2018) for sensory analysis, show highest sensory score was recorded for bananas exposed to 25 g CaC_2 kg^{-1} and lowest sensory score was recorded in fruit that had not been exposed to any chemical for ripening initiation (control samples). Acceptability was found to increase with increasing in CaC_2 concentration. The results for sensory analysis agreed with the finding of Sarananda (1990) who recommend the use of acetylene generated from calcium carbide in ripening of 'Embul' bananas. Amarakoon et al. (1999) also recommended the use of 1 g CaC_2 kg^{-1} to ripen bananas. However, Gunasekara et al. (2015) recorded low sensory scores in 'Embul' bananas treated with either CaC_2 or Ethrel. In tests in Thailand (Jiraporn and Thompson 2019, unpublished results) with Food Technologist Academics mainly from Thailand, but also from India and the UK, bananas from the same bunch of 'Namwa' were initiated to ripen with CaC_2 or left to ripen without any chemical at room temperature of about 30 °C. Their response to the flavour and texture varied with some people preferring one or the other or others said that they could not see any differences. From this it was concluded that the CaC_2 only affected the speed and evenness of ripening but not sensory characteristics.

Toxicity of Calcium Carbide

Industrial grade CaC_2 contains about 80% CaC_2, 15% calcium oxide and 5% other impurities (Kirk-Othmer 2004). Great care must be exercised and the operator must wear protective clothing, including a protective face mask and leave the area immediately after application. Hazard warnings should be put up and flames, cigarettes or electrical fittings that could cause a spark must be eliminated from the area. Industrial grade calcium carbide may contain arsenic and phosphorous hydride (Siddiqui and Dhua 2010). When calcium carbide is applied as a ripening agent, sulfur, arsenic and phosphorous from calcium carbide may diffuse into the peel and flesh of fruits and may pose serious health risk (Islam et al. 2016), which appear to be the main reason for concern about its effect on human health (Chow 1979). Calcium carbide is not "Generally Recognized As Safe" (Pokhrel 2013) and prohibited in several countries including Sri Lanka, India, Bangladesh and Nepal. Sri Lanka under the Section 26 of the Food (Labelling and Miscellaneous) Regulation of 1993. In India use of calcium carbide is strictly banned as per the Prevention of Food Adulteration Act [Section 44AA]. Prevention of Food Adulteration Rules 1955. This prohibits the use of carbide gas for fruit ripening. Food Safety and Standards Regulations 2011 prohibit the selling of artificially ripened fruits using carbide gas. Bangladesh by the Pure Food Ordinance (Amendment) Act 2005, Formation of National Food Safety Advisory Council (NFSAC); prohibit using calcium carbide, formalin, and pesticides in foods. The penal code of Bangladesh must penalize any individual selling illegally ripened fruits. The Nepal Food Regulation 2027, prohibits the use of carbide gas for fruit ripening. In India Chandel et al. (2018) reported that arsine (AsH_3) and phosphine (PH_3) present in CaC_2 as impurities might find entry in the calcium carbide ripened fruits. CaC_2 ripened mangoes contained harmful arsenic levels between 34.73 and 73.43 ppb as compared to high residue levels in open market mangoes (106.27 ppb), while fruits ripened without CaC_2 did not contain detectable residues. Further, dipping fruit in 2% Na_2CO_3 or 2% Agri-biosoft solution for 12 hr. was effective in reducing the arsenic residue from 71.02 ppb to 6.74–9.05 ppb from fruit surfaces and also removed arsenic from peel and pulp although the authors do not report the chemical analysis of Agri-biosoft. However, no information could be found on the residues of arsine and phosphine in the pulp of bananas after being ripened with calcium carbide. Lakade et al. (2018) developed a gold nanoparticle colorimetric method to detect residues of arsenic on fruit surfaces in order to ascertain whether calcium carbide had been used in ripening the fruit.

Harvey (1928) reported "Ethylene oxide, methylene chloride, ethyl chloride, ethylene chloride, propylene chloride, and amylene are desirable on account of their physical property of being liquid at ordinary temperatures at comparatively low pressure, but they were found not as effective as ethylene in ripening fruits, and they all show greater toxicity, shown by the blackening. Amylene, as it is available commercially, has an objectionable odour and produces objectionable flavours in the fruit."

Propylene

Propylene is also called propene or methyl ethylene. Sfakiotakis et al. (1999) showed that endogenous ethylene production in fruit could be induced by exposing them to exogenous propylene. McMurchie et al. (1972) applied propylene to bananas and also found that it increased both their respiration rate and endogenous ethylene production. Golding et al. (1998) successfully initiated pre-climacteric 'Williams' bananas to ripen by exposure them to water saturated air containing 500 mL L^{-1} propylene for 24 h. Harvey (1928) reported "Tasting squads, including horticulturists trained in testing fruits for flavour, have agreed that the flavor of ethylene-treated fruit is superior to that of untreated fruit. Bananas treated with propylene have a little better flavour than those treated with ethylene". Stavroulakis and Sfakiotakis (1997) treated kiwifruit, with 130 µL L^{-1} propylene in a continuous flow-through system at 20 °C in combination with different O$_2$ concentration and found that ethylene biosynthesis and ripening were "stimulated" in the combination of propylene with 21 kPa O$_2$.

Esters

"The esters-amyl acetate, ethyl acetate, and methyl acetate produce browning and blackening of the peel without ripening the fruit. Amyl acetate seems more toxic than methyl or ethyl acetate. The vapor of amyl alcohol was not particularly toxic. Acetaldehyde produced a brown colour similar to that of fruits ripened without gas" (Harvey 1928).

Alcohol

Goonatilake (2008) tested ethylene glycol on bananas and several other fruits and found it could initiate ripening. Their application method was to make up 20% ethylene glycol in water then dip each fruit followed by brushing the solution over the surface. The main impurities found in commercially available ethylene glycol were diethylene glycol (3-oxapentane-1,5-diol), triethylene glycol (3,6-dioxaoctane-1,8-diol), methanol and aldehydic oxidation products (Marcus 1990). Stahler (1962) took out a patent that stated "It has been found that the ripening time of green bananas and citrus fruit can be significantly reduced by maintaining the fruit, after picking and during storage, in contact with a higher alcohol. The alcohols used according to this invention can be any of the alkyl alcohols containing between 6- and 14- carbon atoms, used alone or in combination. For the treatment of green bananas, lauryl alcohol is preferred."

Carbon Monoxide

CO is a colourless, tasteless, odourless and flammable gas with explosive limits in air of between 12.5% and 74.2% by volume. It is extremely toxic to humans and adequate safety precautions must be followed if and when it is used. Plant tissues can oxidize CO to CO_2. CO has been shown to initiate ripening of bananas when they were exposed to concentrations of 0.1% for 24 h at 20°C (Thompson 1996) and to "speed up ripening of tomatoes" (Beckles 2012). CO has other uses in postharvest of fruit and vegetables and can have both beneficial and harmful effects. With levels of O_2 between 2 and 5%, CO can inhibit discoloration of lettuce on the cut butts or from mechanical damage on the leaves. The respiration rate of lettuce was reduced during a 10-day storage period at 2.5 °C when CO was added to the store, but low concentrations of CO in the store atmosphere can increase the levels and intensity of the physiological disorder russet spotting (Table 5.1). Zhang Shaoying et al. (2015) showed that CO ($10 \mu L L^{-1}$) delayed the internal browning of peaches. The control fruit showed internal browning on the 10th day and thereafter the index increased rapidly. CO effectively reduced the development and severity of internal browning by 50%, after 30 days of storage at 8 °C. PAL activities in the CO-treated fruit, showed a trend of first increased and then decreased, the peak appeared on 15th day. Compared with the control, the treatments with CO, inhibited the increase of PPO activities. Firmness of the fruit treated with CO were significantly higher ($p \leq 0.05$) compared to the control during 25 days storage. The treatment with CO, could inhibit the change of PG activity. A combination of 4% O_2, 2% CO_2 and 5% CO was shown to be optimum in delaying ripening and maintaining good quality of mature green tomatoes stored at 12.8 °C after being subsequently ripened at 20 °C (Morris et al. 1981). In peppers and tomatoes the level of chilling injury symptoms could be reduced, but not eliminated when CO was added to the store. CO has fungistatic properties especially when combined with low O_2. *Botrytis cinerea* on strawberries was reduced where the carbon monoxide level was maintained at 5% or higher in the presence of 5% O_2 or lower. Decay in stored mature green tomatoes stored at 12.8 °C was reduced when 5% CO was included in the storage atmosphere (Morris et al. 1981). Tomatoes kept in 5 or 10% (v/v) CO with 4% (v/v) O_2, had superior TSS and total acidity profiles compared to those kept in air with less incidence of *B. cinerea* (Kader 1983).

Figures followed by the same letter were not significantly different ($p = 0.05$).

Table 5.1 The effect of carbon monoxide on the development of russet spotting on harvested lettuce. Modified from Kader 1987

Carbon monoxide concentration $\mu L L^{-1}$	Russet spotting score
0	0a
80	16c
400	22d
2000	24d
10,000	20d
50,000	6b

Smoking

Smoking can produce various gases, depending of the combustion material used to provide the smoke, including acetylene, ethylene and carbon monoxide, all which can initiate fruit ripening. Smoking is used in many developing countries and a study in the Sudan showed that it appeared to be effective, but there was considerable variation in the speed of ripening of the bananas in different parts of the room (Thompson, unpublished results) Kulkarni et al. (2011) reported that in India smoking was used but was unpredictable and could result in uneven ripening and poor colour development. In Sri Lanka a traditional method was described where bananas are laid in a pit covered with banana leaves or a sheet and smoke is generated externally, from burnings semi dried leaves, which is then directed into the pit. Narasimham et al. (1971) compared the traditional practice in India of smoking bunches of bananas for 24–36 h compared with non-smoking on the speed of ripening. They found that smoking reduced the time taken for the peel to change from green to fully yellow from 7 days in non-smoked fruit to 5 days for fruit that had been smoked. The smoking practice they described was "What we do is this: We dig a pit in earth that is enough to put the whole banana cluster in. Then, after safely laying the bananas in the pit, we cover up the pit with a sheet such that only a small hole from a side remains: visualize a small 3–4 inch door to the pit. After that, we light a fire with semi-dry leaves just outside the pit's door. (Semi-dry leaves are used to get as much smoke as possible. Dry leaves do not give that much smoke, because they completely oxidize quickly). And the smoke is sent through the door by blowing it with the aid of a bamboo. This sends a good amount of smoke and warms the inside of the pit considerably. And by experience I can tell you that this makes the bananas to ripen really quickly. I have done a controlled experiment where half of the cluster was not put into the pit. Bananas in the pit ripen overnight and the control sample took days to ripen." Venkata Subbaiah et al. (2013) tested smoking by burning straw, leaves and cow dung in a closed chamber for 24 h and compared this with Ethrel dipping (1000 ppm). Fruit that had been exposed to smoking had a higher proportion of firm fruits after ripening than those that had been initiated to ripen with Ethrel. Mebratie et al. (2015) also compared ripening initiation with Ethrel and smoking. Smoking resulted in faster ripening but led to the least marketability, 29% at 10th day of storage, compared to the other treatments that had more than 83% marketability. The reduction of marketability of smoked bananas was due to blackening and over softening. Gautam and Dhakal (1993) commented that Nepalese farmers have used their indigenous knowledge in ripening of banana since time immemorial. In many places in Nepal banana bunches ('Malbhog') were placed in jute sacks and hung over a fire.

Kerosene

Denny (1927) found that the smoke from kerosene lamps contained ethylene and used the smoke from kerosene lamps to de-green citrus. In some countries, kerosene burners are used to generate smoke in commercial scale banana ripening rooms and

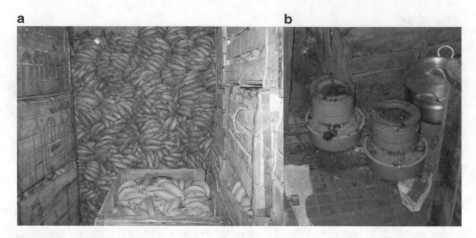

Fig. 5.12 (**a**) Bananas stacked in heaps or wooded boxed in a ripening room in the Souk in Khartoum in Sudan and (**b**) kerosene burners placed in the rooms to initiate ripening in Asmara in Eritrea. Photographs taken (**a**) in 1973 and (**b**) in 2008

a study in the Sudan (Fig. 5.12) showed that it appeared to be effective, but there was considerable variation in the speed of ripening throughout the room, which was assumed to be due to inadequate air circulation. Banana bunches were exposed to smoke generated by burning kerosene stoves inside the chambers for about 24 h. As a result, the temperature inside the chamber also increases besides evolving ethylene gas with traces of other gases like acetylene and carbon monoxide (Ram et al. 1979). Other workers have reported that kerosene fumes also contain other gases including sulphur, aromatic compounds and hydrocarbons (Khan et al. 2013) SO_2, NO, NO_3 and CO (Islam et al. 2016).

Incense

In India Venkata Subbaiah et al. (2013) compared smoking with burning incense sticks and Ethrel dipping in confined rooms on ripening of 'Grand Naine' bananas and found that the incense was ineffective.

Heat

Clearly where fire is used in ripening rooms, along with gases, the temperature will be elevated, often to high levels in tropical countries where it is commonly applied. This high temperature could have an additional effect to the gases. However, Giri et al. (2016) exposed bunches of 'Cavendish' at 40 or 45 °C for 30, 45 or 60 mins and found that none of the heat treatments hastened the speed of ripening compared to the control. The higher the temperature and the longer duration of heat exposure, the slower was the rate of ripening. Kamdee et al. (2018) found that subsequent ripening of 'Sucrier', previously ripened to colour index 3–4, was normal when exposed 42 °C, even for 24 h, but they found their taste and odour was reduced to unacceptable levels. Paull and Chen (2000) studied the extent of the alternation of fruit ripening following heat treatment. They reported that sensitivity or tolerance to heat stress of a fruit or vegetable is related to the level of heat protective proteins at harvest and the postharvest production of heat shock proteins. This sensitivity can be a normal cellular response (less than 42 °C) that can lead to reduced chilling sensitivity, delayed or slowed ripening and a modification of quality or, at greater than 45 °C, the pre-stress environmental conditions can modify the cellular response to stress and cellular recovery.

Damage and Stress

Wounding the banana bunch stalks or fruit may produce ethylene in response to the wound. Generally, any damage to the fruit tissues induces ethylene biosynthesis. Bautista (1990) described a simple technique used in Southeast Asia where a stick is inserted into the stalk of Jackfruit (Fig. 5.13), which not only makes a convenient handle for carrying but damage to the fruit also initiates it to ripen. Bautista also mentions that other methods including cutting, scraping or "pinching" papaya, chico or avocado can hasten ripening.

Ferris et al. (1993) assessed the effects of different types of injury on the speed of ripening of three plantain genotypes in Nigeria and found some hastening of ripening in injured plantains (Table 5.2), but results were not consistent. When a similar experiment was carried out under controlled temperature there were consistent increases in the ripening rate due to damage, but only damage caused by abrasion was consistently significantly ($p = 0.05$) higher than the controls (Fig. 5.14).

Fig. 5.13 Jackfruit in Thailand being initiated to ripen by pushing a stick into the fruit, before sending it to market. Photograph taken in February 2019

Table 5.2 Effects of type of injury on number of days for green plantains, harvested at mature or fully mature, to be ripe. Modified from Ferris et al. (1993)

Harvest maturity	Type of damage			
	None	Impact	Abrasion	Incision
Fully mature	11.0	9.4	9.9	9.8
Mature	16.2	16.9	13.0	15.1

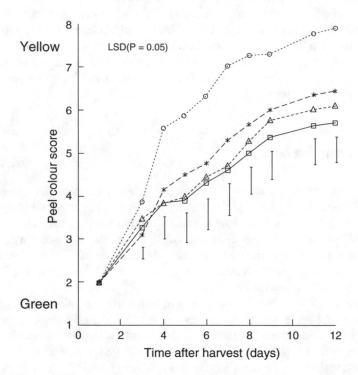

Fig. 5.14 Effects of different types of damage to 'False Horn' plantains on ripening at 20 °C. □ = control, Δ = impact, o = abrasion, x = firmness (static pressure) (Modified from Ferris et al. (1993)

In Brazil Maia et al. (2014) working with 'Dwarf-Prata' bananas showed that all fruits subjected to mechanical injury had increased weight loss, electrolyte leakage, PPO activity, accelerated peel colour changes and an earlier climacteric peak, compared to controls. The damage caused by abrasion caused higher weight loss. The starch conversion to sugars in the pulp was affected by impact damage. The impact and compression damages hastened climacteric ethylene peak and, consequently, fruit ripening. The impact damage greatly increased PPO and POD activities. Stress-induced ethylene biosynthesis was reported by Abeles and Abeles (1972) to be produced by the same biochemical pathway as that produced during the ripening of climacteric fruit. Exposing fruit postharvest to low humidity may also result in sufficient stress to the fruit to initiate ripening. This was shown by Thompson et al. (1974) for plantains. Sfakiotakis et al. (1999) showed that ethylene production was induced autocatalytically by exposure to chilling temperatures, wounding and *Botrytis* infection in Kiwifruit.

In a comparison of transporting 'Dwarf Cavendish' bananas from the field to ripening rooms in the Sudan, Thompson and Silvis (1975) found that dehanding and packing the hands in wooden boxes for transport compared with transporting them as bunches in the traditional way resulted in the ones that had been transported as hands ripening in 22 days compared to those transported as bunches ripening in 18 days, both at 19–23 °C. This difference was significant ($p = 0.05$). It was thought that this shortening of the pre-climacteric period could have been due to the greater damage inflicted on bunches during transport compared to hands packed carefully in boxes. Also, because of the lack of transport in rural areas in those days people would often travel sitting on top of the pile of bunches of bananas, causing more damage.

Fruit Generation

Fruit that are ripening, and thus giving out ethylene, can be placed in an airtight room with green fruit. A continuous system could be worked out for commercial application of this method. However, the room would need to be frequently venti-lated to ensure there was no build-up of CO_2, which is known to inhibit the effect of ethylene. Kitinoja and Kader (2002) commented that 'Small-scale wholesalers and retailers can ripen fruits in bins or large cartons by placing a small quantity of ethylene-generating produce such as ripe bananas in with the produce to be ripened. Cover the bin or carton with a plastic sheet for 24 h, then remove the plastic cover'. Small trials were carried out by A K Thompson (unpublished) in the 1970s in the Sudan with bananas for the local market, but the persons operating the ripening rooms found that using bananas that had begun to ripen to initiate those that had not to be insufficiently predictable and preferred other methods. A study of fruit being used to initiate ripening in bananas was carried out in India by Gandhi et al. (2016). They ripened bananas at ambient temperature in either paper bags or plastic bags in which were included calcium carbide, an apple, a pear or a tomato. There were 3 bananas in each bag. All the fruit in paper bags took 10 days to ripen, as did the control in a plastic bag. However, those with calcium carbide ripened in 5 days (for both 1 and 2 g) while those with an apple ripened in 4 days. Those with a pear rip-ened in 7 days and those with a tomato in 9 days. Tigist Nardos Tadesse (2014) stored 'Dwarf Cavendish' bananas with avocado fruits and found that the ripening period of the bananas was reduced without any undesirable effect on their quality. Pokhrel (2013) also recommended this method and suggested that ripe fruits can be placed with unripe mature fruits in a ratio from 1:20 in an open environment and in a ratio of 1:100 in a closed environment to initiate ripening. Ram et al. (2009) mixed ripe bananas with green bananas and found that the green bananas began to ripen in 3 days and were ready for use in just over 4 days (with an excellent flavour) com-pared to control fruit that did not begin to ripen until 9 days. Burg (2004) described an experiment where bananas and tomatoes were stored at 14.4 °C, either together or separately, either under 760 or 80 mm Hg hypobaric pressure. The bananas stored with tomatoes under 760 mm Hg initiated to ripen because of the ethylene given out by the tomatoes while the bananas under 80 mm Hg did not initiate to ripen.

Leaves

A one-day treatment of bananas with rain tree leaves (*Gliricidia* sp.) at 5% of fruit weight enhanced ripening of 'Latundan' (*Musa* ABB) and 'Saba' (*Musa* BBB) bananas but was less effective than calcium carbide treatment (Acedo and Bautista 1993). Sogo-Temi et al. (2014) tested exposing fruit to freshly harvested leaves from two trees, *Irvingia gabonesis* and *Jathropha curcas*, and compared their effect with calcium carbide and potash (there was no definition of exactly the form of potash used). The calcium carbide and potash were wrapped in polyethene film before being dropped into a black polythene bag containing the bananas. The *I. gabonesis* and the *J. curcas* leaves were washed in cold water and also placed in polythene bags containing bananas, while others were placed in the polyethene bags without any ripening agent and used as the control. The order of ripening of the bananas was: calcium carbide (3 days) then potash, *I. gabonesis* and *J. curcas* and finally the fruit with no ripening agent ripening in 6 days. Pokhrel (2013) reported that in some parts of Nepal, Asuro and Dhurseli (*Colebrookea oppositifolia*) leaves and fresh rice straw, are also used to ripen bananas using about ten times the amount of bananas than the amount of leaves and leaving them covered for a week to ripen. Ram et al. (2009) included dried leaves of Asuro (*Adhatoda vesical* or *Justicia adhatada*) and Koiralo (*Bahunia veriagata*) in sealed PE bags with bananas (100 g each per kg of fruits) and found that the bananas began to soften in 4 days, or just over 4 days, compared to 9 days for controls, and were suitable for consumption after 6 days. They were considered to have a good flavour. da Costa Nascimento et al. (2019) ripened 'Pacovan' bananas using *Bowdichia virgilioides* leaves, compared to calcium carbide, following the empirical method used by Borborema farmers in Brazil. They found that bananas exposed to *B. virgilioides* leaves had a higher respiration rate and ascorbic acid production and reduced acidity and chlorophyll compared to those not treated and there were no significant differences from the bananas that had been initiated to ripen with calcium carbide.

Chapter 6
Ripening Technology

Ripening Rooms

Ripening rooms are designed for climacteric fruit that are harvested before they are ripe and placed under controlled conditions in order to initiate and control ripening. Since the 1990s there has been an increasing demand, in many countries, for all the bananas being offered for sale in a supermarket to be at the same stage of ripeness so that it has an acceptable and predictable shelf-life. This led to the development of a system called "pressure ripening". The system involves the direction of the circulating air in the ripening room being channelled through boxes of fruit so that exogenous ethylene is in contact equally with all the fruit in the room. At the same time the CO_2, which can impede ripening initiation, is not allowed to concentrate around the fruit (Fig. 6.1). In pressurized ripening rooms air is forced though each pallet or series of pallets before returning to the evaporator. Therefore, any "air-stacking" or "cross-stacking" of boxes is not necessary, and the result is less handling of the fruit and improved product quality. In some designs, inflatable bags are placed between pallets to help direct the air through the boxes of bananas. For non-pressurized rooms, the boxes of bananas should be "air stacked". That is, the boxes should be offset to allow the air to circulate among all the boxes since a non-pressurized room design may not facilitate the circulating air to pass through boxes but around them, since the air will take the path of least resistance. In summary the primary requirements for ripening rooms are that they should have:

good temperature control
good and effective air circulation
a good system for introducing fresh air
and be gas tight

Before the introduction of pressure ripening, boxes of bananas were removed from their pallets and stacked in the ripening room so that here was space around

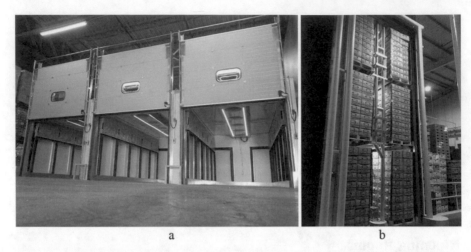

a   b

Fig. 6.1 Pressure ripening rooms (**a**) and pallets of bananas loaded into a pressure ripening room (**b**). Reproduced with permission of Dave Rodden, Advanced ripening technologies Ltd. UK

each box in order to facilitate air circulation (Fig. 6.2). In international trade the boxes of bananas are lined with polyethylene film (Fig. 6.3) and usually transported to ripening rooms stacked on pallets. Several systems have been used. One is to remove each box from the pallet, pull back the plastic film and re-stack them on pallets so that there is a space between boxes (Fig. 6.2). In other cases, the pallets are stacked so that one hand hole in each box is facing outwards. As the fruit are loaded into the rooms the plastic just inside each hand-hold is torn to facilitate air exchange. This is especially important where fruit have been vacuum-packed.

Air circulation systems are usually largely convectional. Air is blown through the cooler and then across the top of the store just below the ceiling. The cooled air falls by convection through the boxes of fruit and is taken up at floor level for recirculation. Many modern ripening rooms have air channels in the floor through which air is circulated at high pressure. This forces it up through the pallets of boxes and should give better air circulation. Special devices such as inflatable air bags placed between pallets are now used to ensure better air circulation and, therefore, more even fruit ripening.

Good ventilation to enable fresh air to be introduced is very important for successful fruit ripening. During the period of initiation of ripening, which is usually 24 h, no fresh air is introduced to the rooms. This is the period when ethylene is introduced. Directly after this period the rooms must be thoroughly ventilated. Setting up a ripening system for bananas in Ecuador, Thompson and Seymour (1984) found that the CO_2 level of ripening rooms had gone up from about zero to 7 kPa during this initial 24 h initiation period. Even with a good fan extraction system it took 40 min of ventilation to bring the CO_2 levels to below 1 kPa. This ventilation with fresh air must be repeated every 24 h during subsequent ripening to prevent levels of CO_2 becoming too high. If rooms are not frequently ventilated ripening can be delayed, or abnormal ripening can occur.

Fig. 6.2 Commercial
banana ripening room in
the 1970s

The rooms need to be gas tight in order to ensure that threshold levels of ethylene are maintained around the fruit during the initiation period. The most common place where leakage occurs is around the doors. It is, therefore, crucial that special gas tight doors are fitted and that they have suitable rubber gaskets. These must be inspected regularly to ensure that they have not been damaged. Gas can also be lost through the walls of ripening rooms. Commonly these are metal-lined on the inside with mastic between joints to ensure no gas can pass through them. Gas tight paint can be used on walls. All holes through the walls for plumbing and electrical fittings must be blocked with mastic. Safety aspects of rooms have already been mentioned. All electric fittings and thermostats must be of a special 'spark-free' type.

It is advisable to have high humidity of 90–95% RH in ripening rooms. To this end many rooms are fitted with some humidification device such as a spinning disc humidifier. However, if the rooms are full of fruit and the refrigerant in the cooling coils used to maintain the room temperature is regulated to within a few degrees of the required room temperature then this should be sufficient to keep the humidity high.

Fig. 6.3 'Cavendish' bananas, on arrval in Europe, in polyethylene film bags that will give a modified atmosphere around the fruit

Catalytic Generators (Fig. 5.4) recommended, for ripening rooms, the following "Banana ripening rooms are very important, not just any room will suffice. A proper ripening room must have the following:

The room must be as air tight as possible to prevent too much of the ethylene from leaking out.

The room must be properly insulated to be able to control the temperature within a few degrees.

The room must have adequate refrigeration. Bananas produce large quantities of heat when they are ripening. The refrigeration equipment must have the capacity to accurately control the pulp temperature.

The room may need heating equipment in order to maintain proper room temperature in cold weather. Electric heating elements have proven the most satisfactory and are often a part of the cooling system. Open flame type heating should never be used.

The room must have adequate air circulation. Because uniform pulp temperatures throughout the load are essential for even ripening, the refrigerated air in the room must circulate at all times and uniformly throughout the load. The room should be constructed so that the air flow path from the refrigeration system, through the load and back to the refrigeration system is unobstructed. Proper air flow patterns are of the utmost importance."

Catalytic Generators also recommend to apply ethylene for a minimum of 24 h during the initial phase of the ripening cycle at 100–150 ppm. To achieve this, the generator setting will depend on the size of the ripening room for example:

Setting 1 for rooms 1600–2500 cubic feet.
Setting 2 for rooms 2500–5000 cubic feet.
Setting 3 for rooms 5000–7500 cubic feet.
Setting 4 for rooms 7500–10,000 + cubic feet.

CO_2 concentrations above 1 kPa can retard ripening, delay the effects of ethylene and cause quality problems. Therefore, it is recommended to ventilate rooms, after the first 24 h of ripening, by opening the doors for 20 min every 12 h. Other venting methods are by automatic fan (either timed or sensor-based) or "flow-though" (constant) ventilation.

An example of the current ripening practices of one major company in UK who ripen 600 to 800 pallets daily was given by Bateman (2019). Each pallet is given a bar code on arrival. The number from the farm that produced the fruit is written on each box enabling feedback to growers to ensure quality control. It takes about 6 days from arrival in the ripening facility to despatch to the retailer but it may take less time depending on growing conditions. On arrival, pallets are tightly packed together and air is directed through each banana box via the handholds and ventilation holes. On first day, the fruit is allowed to come to the correct temperature for ripening (16 °C) as opposed to the transport temperature of 13–14 °C. Then on the second day, ethylene is introduced using catalytic generators. There is a heater and refrigeration unit at the top of each ripening room and the air is forced down. On day three the bananas are heated to about 18–19 °C and on day four or five they are cooled to 13.8–14 °C.

Modelling

Toemmers et al. (2010) described a dynamic model of banana ripening using the conversion of starch to sugars and based on CO_2 emissions. The model was tested on ripening bananas and was shown to describe the characteristic changes in the CO_2 concentration in the air of the ripening facility and the starch concentration in the pulp and the peel colour of the bananas. Based on this information, it was possible, using the model, to estimate the stage of ripeness of the bananas. The model was also used for automated predictive and adaptive process control in banana ripening facilities using an "Open-Loop-Feedback-Optimal" controller. The starch level, the CO_2 reaction rate and the peel colour change rate were fitted to changing process courses, which enabled the control function of the banana ripening. The set-point temperature profile was optimised to automatically control the process to the required ripening stage and the optimal storage temperature at the end of the ripening procedure.

Seo and Hosokawa (1983a) and Seo and Hosokawa (1983b), evaluating automation of banana ripening, confirmed, what had been previously shown, that sugar content of bananas increased during ripening initiation and this was closely correlated with the cumulative quantity of evolved CO_2 by respiration. Thus, a sugar content of bananas at any moment during ripening could be predicted by knowing the cumulative quantity of CO_2 evolved from bananas and that the cumulative quantity of evolved CO_2 at any moment of ripening process could be calculated by CO_2 evolution rate curves at the particular temperature schedule. Conversely, it would be possible to construct a ripening temperature schedule so as to achieve a given value of cumulative quantity of evolved CO_2 or a sugar content of bananas at the termination of the ripening process. Amano et al. (1993) tried to manage cumulative quantity of evolved CO_2 by a conventional control algorithm to control the banana ripening process, but the stable control of cumulative quantity of evolved CO_2 was not always carried out. With fuzzy control, the experience and the knowledge of skilled operators, usually expressed qualitatively in words, are applied to control processes by putting them into fuzzy control rules. Therefore, Seo et al. (1995) considered fuzzy control to be appropriate to apply for automation of banana ripening. Fuzzy logic is widely used in machine control and is based on a mathematical system that analyses analogue input values in terms of logical variables that take on continuous values between 0 and 1. Fuzzy control was applied by Seo et al. (1995) to bananas, which had been initiated to ripen with ethylene, over a 7 day period with the fuzzy rules of three inputs and one output. The operating conditions for ventilation was based on CO_2 concentrations and the deviation of cumulative quantity of evolved CO_2 from the target value. These were considered to reflect the effects of CO_2 and O_2 concentrations in the ripening chamber on the CO_2 evolution rate of bananas. Changes in the measured value of cumulative quantity of evolved CO_2 in ripening by fuzzy control, closely followed changes in the target value. Ventilating operations were effectively conducted to improve the control. Results appeared encouraging with 6050 mg kg^{-1} of cumulative quantity of evolved CO_2 and 10.1% increase in TSS content at the end of ripening against the target values of 6030 mg kg^{-1} CO_2 and 11.8 TSS.

Transport

International trade in bananas was first reported to be 500 stems to from Jamaica to Boston USA in 1866, which took 14 days and the fruit were said to have been sold at a profit (Sealy et al 1984). There was no indication of temperature control during this or subsequent shipments in the latter part of the nineteenth century. In 1901 the vessel Port Morant carried 23,000 stems of bananas from Jamaica to Britain using mechanical refrigerated holds for the company Elder and Fyffes (Sinclair 1988).

Green banana transport is usually by sea freight at 13-14 °C, 85–95% RH and 25–60 cm of water (cmH) ventilation for about 18–22 days. Using controlled atmospheres for transporting bananas internationally, in order to prevent ripening initiation, has been used for many years in reefer containers and reefer ships. See section on controlled atmosphere storage (Chap. 4). Janssen (2014) and Jedermann et al. (2015) tested the construction of a container prototype that was capable of remote supervision of the physiological stage of fruit. It had a sensor system comprising of a set of wireless temperature and humidity sensors to detect local maximum temperatures and hot spots as well as sensors for CO_2 and ethylene. Tests were carried out using a system for automated transport supervision on commercial shipments of bananas that focused on the detection of "ship ripes" by their higher respiration rate and ethylene production as well as by the detection of potential hot spots caused by higher respiration resulting in heat production combined with insufficient air flow. These tests were carried out during transport of bananas from Costa Rica to Europe using a simple heat transfer model that allowed estimation of index values for local cooling effects and respiration rate in each pallet from the measured temperature curves. They found that the model could predict the risk of hot spots, but the major obstacle was the large amount of heat generated that was addressed by using different modifications in the packing scheme to improve the air flow, which showed clear benefits with regard to the amount of heat removed by cooling.

Using controlled atmospheres for transporting bananas internationally has been used for many years. For example, Dr. Errol Reid carried out experiments in 1997 on CA transport of bananas in reefer containers from the Windward Islands to Britain and subsequently supervised the use of CA reefer ships in 1998. CA reefer ships were being used for transport of bananas from the Windward Islands to the UK with some 50,000–60,000 tonnes per year in 2010 often combined with bananas from the Dominican Republic to reduce the risk of dead-freight. Maersk market CA containers (called Starcare™) that they claim will extend the green life of bananas for to up to 50 days (Maersk 2019).

Reducing Ripening Initiation in Transit

For international transport of bananas storage conditions are commonly set at 13.3 °C and 85–95% RH and ventilation at 25 m^3 $hr.^{-1}$, which should give a pre-climacteric life of about 4 weeks. An equation describing the relationship between temperature, ethylene concentration and green life of 'Cavendish' bananas was developed and applied to a shipment protocol of 19 days for bananas exported from Central America to southern Europe by Wills et al. (2014). The equation predicted that fruit could be transported without refrigeration if ethylene levels were maintained at 0.04 μL L^{-1} during the winter temperature of 17 °C and at 0.002 μL L^{-1} at the summer transport temperature of 24 °C.

Hyperbaric conditions, even at variable room temperatures of up to 37 °C, have been shown to preserve foods and thus achieve significant energy savings (Fernandes et al. 2015). Designing shipping containers with the facility of increasing the atmospheric pressure inside each container would be an alternative way of maintaining bananas in a preclimacteric condition during transport. Hyperbaric storage, at room temperature, could be more energy efficient that refrigeration since the only energy costs are during compression and no additional energy is required to subsequently maintain the product under pressure. However, the capital costs of high-pressure equipment are high (Saraiva 2014), but expansion of high-pressure food processing should help to reduce equipment costs (Balasubramaniam et al. 2008). Bartlett pears in storage at 20 °C had a higher ethylene production rate in 100 kPa O_2 than in air (Frenkel 1975). Similar results were reported by Morris (1981) for mature-green and breaker tomatoes stored at 20 °C when they were exposed to 30 or 50 kPa O_2 but exposure to 80 or 100 kPa O_2 reduced ethylene production rates and musk-melons stored in 100 kPa O_2 at 20 °C had similar ethylene production level as those stored in air (Altman and Corey 1987). Zheng et al. (2008) found that zucchini squash exposed to 60 kPa O_2 had a lower ethylene production compared to those stored in 100 kPa O_2 or in air. Hypobaric containers have been tested for international transport of fresh food. While hypobaric conditions have been shown to increase the postharvest life of fresh bananas it has so far not been considered practical and economic for commercial use in banana transport (Burg 2004; Thompson 2016; see also Chap. 4 Hypobaric Conditions).

Ripening in Transit

Ripening rooms are expensive to build and operate. Since bananas cannot be successfully allowed to ripen before harvest and they are often transported over large distances for protracted periods, it seems logical that at least ripening during transport should be considered as an option. Jedermann et al. (2014, 2015) commented that ripening of bananas can be carried out directly in the reefer container, used for transport, after it arrives at its destination. Several companies have developed ripening systems for bananas in containers during actual transit from the field to the retail outlet. University of Queensland (2017) commented, referring particularly to mangoes, that "in-transit ripening is of significant interest to the industry with the opportunity to substantially reduce time to market and minimize the pressure on large and expensive ripening room infrastructure". For in-transit ripening of mangoes using their IC powder (ethylene-apha-cyclodextrin), Ho et al. (2016) found ethylene at between 4.9 and 10.5 µL L^{-1} in the headspace of the containers over 48 h where the IC powder had been used. Mangoes from the treated containers had a shorter ripening time by 3–6 days compared to the non-treated control fruit.

Dole Food adopted a technology that ripens bananas in shipping containers. A briefcase-like metal box containing ethylene canisters is used to release ethylene gas at a controlled rate for shipping times from 1 to 7 days, so that fruit can be at the required stage of ripeness on arrival. In this way fruit can be distributed directly to the retail market without first being taken to a ripening facility.

The American company SmartAir Technology use modified container and refrigeration units for in-transit ripening. Two pods are installed at each side of the front bulkhead of the container with four blowers in two fan pods controlled by a SmartAir control box mounted at the front of the trailer, which also operates the ethylene gas generator and the fresh air exchange system the unit runs continuously when the SmartAir system is in use. Discharge air from the refrigeration unit evaporator empties into the mixing chamber behind the front bulkhead. The discharge mixes with air in the trailer for precise temperature control, typically within ±1 °F at any point in the trailer. Most of the mixed air is captured by the fan pods and directed down the trailer sidewalls instead of flowing above the load as it would in a conventional refrigerated trailer (Macklin 2001).

A patent was taken out for ripening bananas inside a shipping container in 2013 by Axel Moehrke. The patent was summarized as "To ripen bananas during shipping, unripened bananas are first placed in ethylene-permeable containers within a shipping box. The boxes of bananas are arranged onto shipping pallets which are then placed into a shipping container for shipping. The arrangement of the bananas within the reefer container is such that there is adequate airflow around the fruits to ensure stable temperature throughout the shipping and ripening process. While enclosed in the reefer container, during shipping, the bananas are exposed to ethylene gas at a slow rate over a number of days, to ensure slow, even ripening. The bananas may be fully or partially ripe upon removal from the shipping container and are then shipped directly to retail outlets or distribution centers, without the need for additional ripening in traditional ripening rooms." (Moehrke 2014).

Chapter 7
Conclusions

Refrigeration technology is continuously improving including improved energy efficiency. The desire for energy efficiency is a top concern with higher energy savings with further advances in the mechanical equipment of ripening rooms including less heat given off by the mechanical equipment. Technology control of the atmosphere remotely is progressing with the provision of software that allows remote access to control ripening rooms, including adjusting alerts, gas parameters and temperatures through Internet connections.

The maturity of individual banana fingers varies with the climate, environment and weather during growth, but also within a single bunch. So, if bananas are allowed to ripen without artificial initiation there will be variation as to when they reach optimum eating maturity. This may be acceptable or even advantages for a household, but it not acceptable in most commercial production.

There seems no clear consensus on whether bananas taste better if not initiated to ripen artificially. The enormous changes in organoleptic properties as fruit ripen makes it difficult to differentiate determine differences and no clear conclusions can be made from chemical analyses. If the function of ethylene, or any of the other ways of initiating ripening, have any other function that initiating, again is not clear and seems unlikely.

A. K. Thompson et al., *Banana Ripening*, SpringerBriefs in Food, Health, and Nutrition, https://doi.org/10.1007/978-3-030-27739-0_7

References

Abeles, A. L., & Abeles, F. B. (1972). Biochemical pathway of stress-induced ethylene. *Plant Physiology, 50*, 496–498.

Aborisade, A. T., & Ayibiowu, A. F. (2010). Effect of polyethylene thickness, photoperiod and initial stage at harvest on ripening of two tomato (*Lycopersicon esculentum* Mill.) cultivars. *Acta Horticulturae, 858*, 179–184.

Acedo, A. L., & Bautista, O. K. (1993). Optimization of indigenous ripening systems for bananas in the Philippines. *ACIAR Proceedings, 50*, 172–185.

Adams, D. O., & Yang, S. F. (1979). Ethylene biosynthesis: identification of 1-aminocyclopropane-1-carboxylic acid as an intermediate in the conversion of methionine to ethylene. *Proceeding of the National Academy of Sciences of the United States of America, 76*, 170–174.

Adão, R. C., & Glória, M. B. A. (2005). Bioactive amines and carbohydrate changes during ripening of 'Prata' banana (*Musa acuminata* × *M. balbisiana*). *Food Chemistry, 90*, 705–711.

Adeyemi, O. S., & Oladiji, A. T. (2009). Compositional changes in banana (*Musa* spp.) fruits during ripening. *African Journal of Biotechnology, 8*, 858–859.

Aghdam, M. S., Asghari, M., Khorsandi, O., & Mohayeji, M. (2014). Alleviation of postharvest chilling injury of tomato fruit by salicylic acid treatment. *Journal of Food Science and Technology, 51*, 2815–2820.

Ahmad, S., & Thompson, A. K. (2006). Effect of controlled atmosphere storage on ripening and quality of banana fruit. *Journal of Horticultural Science & Biotechnology, 81*, 1021–1024.

Ahmad, S., & Thompson, A. K. (2007). Effect of modified atmosphere storage on the ripening and quality of ripe banana fruit. *Acta Horticulturae, 741*, 273–278.

Ahmad, S., Clarke, B., & Thompson, A. K. (2001a). Banana harvest maturity and fruit position on quality of ripe fruit. *Annals of Applied Biology, 139*, 329–335.

Ahmad, S., Thompson, A. K., Hafiz, I. A., & Asi, A. A. (2001b). Effect of temperature on the ripening behaviour and quality of banana fruit. *International Journal of Biology, 3*, 224–227.

Ahmad, S., Perviez, M. A., Chatha, Z. A., & Thompson, A. K. (2006). Improvement of banana quality in relation to storage humidity, temperature and fruit length. *International Journal of Agriculture and Biology, 8*, 377–380.

Ahmad, S., Thompson, A. K., & Pervez, M. A. (2007). Effect of fruit maturity and position of hands on the ripening behaviour and quality of banana fruit. *Acta Horticulturae, 741*, 117–123.

Ahmed, Z. F. R., & Palta, J. P. (2011). A natural lipid, lysophosphatidylethanolamine, may promote ripening while reducing senescence in banana fruit. *HortScience, 46*, 273.

Ahmed, Z. F. R., & Palta, J. P. (2015). A postharvest dip treatment with lysophophatidylethanol-amine, a natural phospholipid, may retard senescence and improve the shelf life of banana fruit. *HortScience, 50*, 1035–1040.

Ahmed, Z. F. R., & Palta, J. P. (2016). Postharvest dip treatment with a natural lysophospholipid plus soy lecithin extended the shelf life of banana fruit. *Postharvest Biology and Technology, 113*, 58–65.

Akkaravessapong, P., Joyce, D. C., & Turner, D. W. (1992). The relative humidity at which bananas are stored or ripened does not influence their susceptibility to mechanical damage. *Scientia Horticulturae, 52*, 265–268.

Alonso, J. M., Hirayama, T., Roman, G., Nourizadeh, S., & Ecker, J. R. (1999). EIN2, a bifunctional transducer of ethylene and stress responses in *Arabidopsis. Science, 284*, 2148–2152.

Altman, S. A., & Corey, K. A. (1987). Enhanced respiration of muskmelon fruits by pure oxygen and ethylene. *Scientific Horticulture, 31*, 275–281.

Alvindia, D. G., Kobayashi, T., Yaguchi, Y., & Natsuaki, K. T. (2000). Evaluation of cultural and postharvest practices in relation to fruit quality problems in Philippine non chemical bananas. *Japanese Journal of Tropical Agriculture, 44*, 178–185.

Al-Zaemey, A. B. S., Falana, I. B., & Thompson, A. K. (1989). Effects of permeable fruit coatings on the storage life of plantains and bananas. *Aspects of Applied Biology, 20*, 73–80.

Amano, T., Seo, Y., Morishima, H., & Kawagoe, Y. (1993). Studies on ripening of fruits by fuzzy theory (Part I). *Journal of the Japanese Society of Agricultural Machinery and Food Engineers, 55*, 137–142.

Amarakoon, R., Sarananda, K. H., & Illeperuma, D. C. K. (1999). Quality of mangoes as affected by stage of maturity. *Tropical Agricultural Research, 11*, 74–85.

Amaro, A. L., & Almeida, D. P. F. (2013). Lysophosphatidylethanolamine effects on horticultural commodities: A review. *Postharvest Biology and Technology, 78*, 92–102.

Ambuko, J. L., Sekozawa, Y., Sugaya, S., Itoh, F., Nakamura, K., & Gemma, H. (2006). Effect of seasonal variation, cultivar and production system on some postharvest characteristics of the banana. *Acta Horticulturae, 712*, 841–849.

Anhwange, B., Ugye, J. T., & Nyiatagher, T. D. (2009). Chemical composition of *Musa sepientum* (banana) peels. *Electronic Journal of Environmental, Agricultural and Food Chemistry, 8*, 437–442.

Anonymous. (2012). *Banana ripening manual*. Boston: United Fruit Sales Corp.

Anonymous. (2019). Ripelock. https://www.agrofresh.com/technologies/ripelock/. Accessed 7 April 2019.

Anuchai, J., Chumthongwattana, M., Tepsorn, R., & Supapavanich, S. (2018). Efficiency of salicylic immersion using fine-bubble technique on quality of *Musa* AAA fruit during ripening. *International Journal of Agricultural Technology, 14*, 1003–1016.

Apelbaum, A., Aharoni, Y., & Temkin-Gorodeiski, N. (1977). Effects of sub-atmospheric pressure on the ripening processes of banana fruits. *Tropical Agriculture, 54*, 39–46.

Archana, U., & Sivachandiran, S. (2015). Effect of application of gibberellic acid (GA_3) on shelf-life of banana. *International Journal of Research in Agriculture and Food Sciences, 3*, 1–4.

Arora, A., Choudhary, D., Agarwal, G., & Singh, V. P. (2008). Compositional variation in β-carotene content, carbohydrate and antioxidant enzymes in selected banana cultivars. *Institute of Food Science and Technology, 43*, 1913–1921.

Badran, A. M. (1969). *Controlled atmosphere storage of green bananas*. U.S. Patent 17 June 3, 450, 542.

Bagnato, N., Sedgley, M., Barrett, R., & Klieber, A. (2003). Effect of ethanol vacuum infiltration on the ripening of 'Cavendish bananas' cv. Williams. *Postharvest Biology and Technology, 27*, 337–340.

Balasubramaniam, V. M., Farkas, D., & Turek, E. J. (2008). Preserving foods through high-pressure processing. *Food Technology, 62*, 32–38.

Ball, K. L., Green, J. H., & Ap Rees, T. (1991). Glycolysis at the climacteric bananas. *European Journal of Biochemistry, 197*, 265–269.

Bangerth, F. (1984). Changes in sensitivity for ethylene during storage of apple and banana fruits under hypobaric conditions. *Scientia Horticulturae, 24,* 151.

Banks, N. H. (1984). Some Effects of TAL Pro-long coating on ripening bananas. *Journal of Experimental Botany, 35,* 127–137.

Barnell, H. R. (1943). Studies in tropical fruits. Carbohydrate metabolism of the banana fruit during storage at 53 °F. *Annals of Botany New Series, 9,* 1–22.

Barry, C. S., & Giovannoni, J. J. (2007). Ethylene and fruit ripening. *Journal of Plant Growth Regulation, 26,* 143–159.

Baskar, R., Shrisakthi, S., Sathyapriya, B., Shyampriya, R., Nithya, R., & Poongodi, P. (2011). Antioxidant potential of peel extracts of banana varieties (*Musa sapientum*). *Food and Nutrition Sciences, 2,* 1128–1133.

Bateman, D. (2019). S.H. Pratt & Co www.freshfruitportal.com. Accessed 26 April 2019.

Bautista, O. K. (1990). *Postharvest technology for southeast Asian perishable crops.* Technology and Livelihood Resource Center 302 pp.

Beatrice, E., Deborah, N., & Guy, B. (2015). Provitamin A carotenoid content of unripe and ripe banana cultivars for potential adoption in eastern. *African Journal of Food Composition and Analysis, 43,* 1–6.

Beckles, D. M. (2012). Factors affecting the postharvest soluble solids and sugar content of tomato (*Solanum lycopersicum* L.) fruit. *Postharvest Biology and Technology, 63,* 129–140.

Belitz, H. D., Grosch, W., & Schierberle, P. (2009). *Food chemistry* (4th ed.). Springer Publications.

Bennett, R. N., Shiga, T. M., Hassimotto, N. M., Rosa, E. A., Lajolo, F. M., & Cordenunsi, B. R. (2010). Phenolics and antioxidant properties of fruit pulp and cell wall fractions of postharvest banana (*Musa acuminata* Juss.) cultivars. *Journal of Agricultural and Food Chemistry, 58,* 7991–8003.

Bhardwaj, C. L., Jones, H. F., & Smith, I. H. (1984). A study of the migration of externally applied sucrose esters of fatty acids through the skins of banana, apple and pear fruits. *Journal of the Science of Food and Agriculture, 35,* 322–331.

Bhova, H. P., Patel, J. C., & Amin, H. D. (1978). Effect of Ethrel on ripening of some varieties of mango (*Mangifera indica* L.) fruits. *Indian Journal of Agricultural Research, 12,* 263–265.

Biale, J. B. (1960). Respiration of fruits. *Encyclopaedia of Plant Physiology, 12,* 536–592.

Biale, J. B. (1964). Growth, maturation and senescence in fruits. *Science, 146,* 880–888.

Biale, J. B., & Young, R. E. (1962). The biochemistry of fruit maturation. *Endeavour, 21,* 164–174.

Blackbourn, H. D., Jeger, M. J., John, P., & Thompson, A. K. (1990). Inhibition of degreening in the peel of bananas ripened at tropical temperatures, III changes in plastid ultrastructure and chlorophyll-protein complexes accompanying ripening in bananas and plantains. *Annals of Applied Biology, 117,* 147–161.

Blackbourn, H. D., Jeger, M. J., John, P., Telfer, A., & Barber, J. (1990a). Inhibition of degreening in the peel of bananas ripened at tropical temperatures. IV. Phytosynthetic capacity of ripening bananas and plantains in relation to changes in the lipid composition of ripening banana peel. *Annals of Applied Biology, 117,* 163–174.

Blissett, K. A., Emanuel, M., & Barnaby, A. G. (2019). Biochemical properties of tree ripened and post harvest ripened *Mangifera indica* (cv. East Indian). *International Journal of Fruit Science,* 1–13.

Bouzayen, M., Latché, A., Nath, P., & Pech, J. C. (2010). Mechanism of fruit ripening. Chapter 16. In *Plant developmental biology – biotechnological perspectives* (Vol. 1). Springer.

Bowden, A. P. (1993). Modified atmosphere packaging of 'Cavendish' and 'Apple' bananas. *MSc thesis Cranfield University.*

Bowden, A. P., Khanbari, O., Wei, Y., & Thompson, A. K. (1994). Implications of genetic variation on the marketing of fruit and vegetables. *Aspects of Applied Biology, 39,* 103–110.

Brady, C. J. (1987). Fruit ripening. *Annual Review of Plant Physiology, 38,* 155–178.

Brat, P., Yahia, A., Chillet, M., Bugaud, C., Bakry, F., Reynes, M., & Brillouet, J. M. (2004). Influence of cultivar, growth altitude and maturity stage on banana volatile compound composition. *Fruits, 59,* 75–82.

Brodrick, H. T., & Strydom, G. J. (1984). The radurisation of bananas under commercial conditions – Part 1. *Subtropical Fruit Journal, 602*, 4–6.

Broughton, W. J., Chan, B. E., & Kho, H. L. (1978). Maturation of Malaysian fruits. II. Storage conditions and ripening of banana *Musa sapientum* var 'Pisang Emas'. *Malaysian Agricultural Research and Development Institute. Research Bulletin, 7*, 28–37.

Bugaud, C., Alter, P., Daribo, M. O., & Brillouet, J. M. (2009). Comparison of the physicochemical characteristics of a new triploid banana hybrid, FLHORBAN920, and the Cavendish variety. *Journal of the Science of Food and Agriculture, 89*, 407–413.

Bugaud, C., Ocrisse, G., Salmon, F., & Rinaldo, D. (2014). Bruise susceptibility of banana peel in relation to genotype and post-climacteric storage conditions. *Postharvest Biology and Technology, 87*, 113–119.

Burdon, J. N., Dori, S., Lomaniec, E., Marinansky, R., & Pesis, E. (1994). The postharvest ripening of water stressed banana fruits. *Journal of Horticulture Science, 69*, 799–804.

Burg, S. P. (1967). *Method for storing fruit*. US Patent3.333967 and US patent reissue Re. 28,995 (1976).

Burg, S. P. (2004). *Postharvest physiology and hypobaric storage of fresh produce*. Wallingford: CAB International.

Burg, S. P. (2014). *Hypobaric storage in food industry*. London: Academic.

Burg, S. P., & Burg, E. A. (1962). Role of ethylene in fruit ripening. *Plant Physiology, 37*, 179–189.

Burg, S. P., & Burg, E. A. (1965). Relationship between ethylene production and ripening in bananas. *Botanical Gazette, 126*, 200–204.

Burg, S. P., & Burg, E. A. (1967). Molecular requirements for the biological activity of ethylene. *Plant Physiology, 42*, 144–152.

Butler, W., Cook, L., & Vayda, M. E. (1990). Hypoxic stress inhibits multiple aspects of the potato tuber wound responses. *Plant Physiology, 93*, 264–270.

Calberto, G., Staver, C., & Siles, P. (2015). An assessment of global banana production and suitability under climate change scenarios. In A. Elbehri (Ed.), *Climate change and food systems. Global assessments and implications for food security and trade* (pp. 264–291). Rome: Food and Agriculture Organization of the United Nations.

Campbell, C. W., & Malo, S. E. (1969). The effect of 2-chloroethyl phosphonic acid on ripening of mango fruits. *Proceedings of the American Society for Horticultural Science, 13*, 221–226.

Cano, M. P., de Ancos, B., Matallana, M. C., Cámara, M., Reglero, G., & Tabera, J. (1997). Differences among Spanish and Latin-American banana cultivars: Morphological, chemical and sensory characteristics. *Food Chemistry, 59*, 411–419.

Caussiol, L. P., & Joyce, D. C. (2004). Characteristics of banana fruit from nearby organic versus conventional plantations: A case study. *Journal of Horticultural Science & Biotechnology, 79*, 678–682.

Chamara, D., Illeperuma, K., & Galappatty, P. T. (2000). Effect of modified atmosphere and ethylene absorbers on extension of storage life of 'Kolikuttu' banana at ambient temperature. *Fruits, 55*, 381–388.

Chandel, R., Sharma, P. C., & Gupta, A. (2018). Method for detection and removal of arsenic residues in calcium carbide ripened mangoes. *Journal of Food Processing and Preservation, 42*, e13420.

Chang, W. H., & Hwang, Y. J. (1990a). Effect of ethylene treatment on the ripening, polyphenoloxydase activity and water-soluble tannin content of Taiwan northern banana at different maturity stages and the stability of banana polyphenoloxydase. Tropical and subtropical fruits. *Acta Horticulturae, 275*, 603–610.

Chang, W. H., & Hwang, Y. J. (1990b). Effect of some inhibitors on carbohydrate content and related enzyme activity during ripening of Taiwan northern banana fruit. Tropical and subtropical fruits. *Acta Horticulturae, 275*, 611–619.

Chang, W. H., Hwang, Y. J., & Wei, T. C. (1990). Chemical composition and enzyme activity of Taiwan northern banana fruit of different maturity and harvested in different seasons. Tropical and Subtropical fruits. *Acta Horticulturae, 275*, 621–629.

Chauhan, O. P., Raju, P. S., Dasgupta, D. K., & Bawa, A. S. (2006). Modified atmosphere packaging of banana (cv. Pachbale) with ethylene, carbon di-oxide and moisture scrubbers and effect on its ripening behaviour. *American Journal of Food Technology, 1*, 179–189.

Cheesman, E. E. (1947–1949). Classification of the bananas. *Kew Bulletin* 1947, 106–117. 1948, 11–28, 145–157, 323–328, 1949, 23–28, 133–137, 265–272, 445–449.

Cheng, G., Yang, E., Lu, W., Jia, Y., Jiang, Y., & Duan, X. (2009). Effect of nitric oxide on ethylene synthesis and softening of banana fruit slice during ripening. *Journal of Agricultural and Food Chemistry, 57*, 5799–5804.

Chillet, M., & De Lapeyre De Bellaire, L. (2002). Variability in the production of wound ethylene in bananas from the French West Indies. *Scientia Horticulturae, 96*, 127–137.

Chillet, M., de Lapeyre de Bellaire, L., Huber, B., & Mbéguié-A- Mbéguié, D. (2008). Measurement of banana green life. *Fruits, 63*, 125–127.

Chotikakham, S., Chumyam, A., & Saengnil, K. (2014). Changes in membrane lipid peroxidation during fruit ripening of 'Namwa' banana. *Agricultural Science Journal, 45*, 113–116.

Choudhury, H., Chandra, K., & Baruah, K. (1996). Variation in total chlorophyll content and its partitioning in Dwarf Cavendish bananas as influenced by bunch cover treatments. *Crop Research, 11*, 232–238.

Chow, M. (1979). The preoccupation with food safety. In M. Chow & T. D. P. Harmon Jr. (Eds.), *Critical food issues for the eighties* (pp. 14–42). New York: Pergamon Press.

Christelová, P., Valarik, M., Hřibová, E., De Langhe, E., & Doležel, J. (2011). A multigene sequence-based phylogeny of the Musaceae (banana) family. *BMC Evolutionary Biology, 11*, 103–116.

Christelová, P., De Langhe, E., Hřibová, E., Čížková, J., Sardos, J., Hušáková, M., Van den houwe, I., Sutanto, A., Kepler, A. K., Swennen, R., Roux, N., & Doležel, J. (2017). Molecular and cytological characterization of the global *Musa* germplasm collection provides insights into the treasure of banana diversity. *Biodiversity and Conservation, 26*, 801–824.

Collin, M. N. (1989). Conservation de bananes plantains sous film plastique et polyolefines. *IRFA Reunion Annuelle, 37*, 7.

Collin, M. N., & Dalnic, R. (1991). Evolution de quelques criteres physico-chimiques de la banane plantain (cultivar Orishele) au cours de la maturation. *Fruits, 46*, 13–17.

Collin, M. N., & Folliot, M. (1990). Caracteristiques anatomiques de l'epiderme de la banane plantain en relation avec les techniques de conservation. *Fruits, 45*, 9–16.

Copisarow, M. (1935). The metabolism of fruit and vegetables in relation to their preservation. *Journal of Pomology, XIV*, 9–18.

Cordenunsi, B. R., & Lajolo, F. M. (1995). Starch breakdown during banana ripening: Sucrose synthase and sucrose phosphate synthase. *Journal of Agricultural and Food Chemistry, 43*, 347–351.

Cuadrado, Y., Fernandez, M., Recio, E., Aparicio, J. F., & Martin, J. F. (2004). Characterization of the *ask-asd* operon in aminoethoxyvinylglycine-producing *Streptomyces* sp. NRRL 5331. *Applied Microbiology and Biotechnology, 64*, 228–236.

Curry, E. A. (1998). *Ethylene in fruit ripening*. 14th Annual Postharvest Conference, Yakima.

da Costa Nascimento, R., de Oliveira Freire, O., Ribeiro, L. S., Araújo, M. B., Finger, F. L., Soares, M. A., Wilcken, C. F., Zanuncio, J. C., & Ribeiro, W. S. (2019). Ripening of bananas using *Bowdichia virgilioides* Kunth leaves. *Scientific Reports* 9, 1–6. Article number: 3548. https://doi.org/10.1038/s41598-019-40053-3

Daniells, J., Jenny, C., Karamura, D., & Tomekpe, K. (Eds.). (2001). *Musalogue: t of controlled atmosphere storage on ripening and quality of banan Musa. A catalogue of Musa germplasm.* Rome: International Plant Genetic Resources Institute (IPGRI).

Daundasekera, W. A. M., Joyce, D. C., Adkikaram, N. K. B., & Terry, L. A. (2008). Pathogen-produced ethylene and the *Colletotrichum musae*-banana fruit pathosystem. *Australasian Plant Pathology, 37*, 448–453.

Davey, M. W., Van den Bergh, I., Markham, R., Swennen, R., & Keulemans, J. (2009). Genetic variability in *Musa* fruit provitamin A carotenoids, lutein and mineral micronutrient contents. *Food Chemistry, 115*, 806–813.

Davies, P. N. (1990). *Fyffes and the banana*. London: Athlone Press.

Davies, K., Hobson, G. E., & Grierson, D. (2006). Silver ions inhibit the ethylene-stimulated production of ripening-related mRNAs in tomato. *Plant Cell and Environment, 11*, 729–738.

De Martino, G., Mencarelli, F., & Golding, J. B. (2007). Preliminary investigation into the uneven ripening of banana (*Musa* sp.) peel. *New Zealand Journal of Crop and Horticultural Science, 35*, 193–199.

Deaquiz, Y. A., Álvarez-Herrera, J., & Fischer, G. (2014). Etileno y 1-MCP afectan el comportamiento poscosecha de frutos de pitahaya amarilla (*Selenicereus megalanthus* Haw.). *Agronomía Colombiana, 32*, 44–51.

Deekshika, B., Praveena Lakshmi, B., Singuluri, H., & Sukumaran, M. K. (2015). Estimation of ascorbic acid content in fruits & vegetables from Hyderabad, India – A theoretical assessment of Vitamin C activity. *International Journal of Current Microbiology and Applied Sciences, 4*, 96–99.

Deepthi, V. P. (2016). Taxonomic scoring and genomic grouping in bananas. *Flora and Fauna, 22, 151–158*.

Deng, Z., Jung, J., Simonsen, J., & Zhao, Y. (2017). Cellulose nanomaterials emulsion coatings for controlling physiological activity, modifying surface morphology, and enhancing storability of postharvest bananas (*Musa acuminate*). *Food Chemistry, 232*, 359–368.

Denny, F. E. (1923). *Methods of colouring citrus fruits*. US Patent no.1, 475,938.

Desai, B. B., & Deshpande, P. B. (1975). Chemical transformations in three varieties of banana *Musa paradisica* Linn fruits stored at 20 °C. *Mysore Journal of Agricultural Science, 9*, 634–643.

Dhembare, A. J. (2013). Bitter truth about fruit with reference to artificial ripener. *Archives of Applied Science Research, 5*, 45–54.

Dole. (2011). Dole offers on-board banana ripening. *The Packer* May 17 2011. https://www.the-packer.com/article/dole-offers-board-banana-ripening. Accessed 21 March 2019.

Domínguez, M., Domínguez-Puigjaner, E., Saladié, M., & Vendrell, M. (1998). Effect of inhibitors of ethylene biosynthesis and action on ripening of bananas. *Acta Horticulturae, 490*, 519–528.

Dominguez-Puigjaner, E., Vendrell, M., & Dolors Ludevid, M. (1992). Differential protein accumulation in banana fruit during ripening. *Plant Physiology, 98*, 157–162.

Duan, X., Joyce, D. C., & Jiang, Y. (2007). Postharvest biology and handling of banana fruit. *Fresh Produce, 1*, 140–152.

Dubery, I. A., van Rensburg, L. J., & Schabort, J. C. (1984). Malic enzyme activity and related Biochemical aspects during ripening of γ-irradiated mango fruit. *Phytochemistry, 23*, 1383–1386.

Elitzur, T., Vrebalov, J., Giovannoni, J. J., Goldschmidt, E. E., & Friedman, H. (2010). The regulation of MADS-box gene expression during ripening of banana and their regulatory interaction with ethylene. *Journal of Experimental Botany, 61*, 1523–1535.

Elitzur, T., Yakir, E., Quansah, L., Zhangjun, F., Vrebalov, J., Khayat, E., Giovannoni, J. J., & Friedman, H. (2016). Banana *MaMADS* transcription factors are necessary for 14 fruit ripening and molecular tools to promote shelf-life and food security. *Plant Physiology, 171*, 380–391.

Embrapa. (2014). Sistema de Produção da Bananeira Irrigada: Cultivares. *Embrapa Semiárido*http://sistemasdeproducao.cnptia.embrapa.br/FontesHTML/Banana/BananeiraIrrigada/cultivares.htm. Accessed 2 July 2015.

Englberger, L., Wills, R. B., Blades, B., Dufficy, L., Daniells, J. W., & Coyne, T. (2006). Carotenoid content and flesh color of selected banana cultivars growing in Australia. *Food and Nutrition Bulletin, 27*, 281–291.

EPA, U.S. (1988). *Guidance for the reregistration of pesticide products containing ethephon as the active ingredient*. Washington, DC: U.S. Environmental Protection Agency.

Esguerra, E. B., Hilario, D. C. R., & Absulio, W. L. (2009). Control of finger drop in 'Latundan' banana (*Musa acuminata* AA group) with preharvest calcium spray. *Acta Horticulturae, 837*, 167–170.

FAOStat. (2019). http://www.fao.org/faostat/en/#data. Accessed 5 May 2019.

Farag, K. M., & Palta, J. P. (1993). Use of lysophosphatidylethanolamine, a natural lipid to retard tomato leaf and fruit senescence. *Physiologia Plantarum, 87*, 515–524.

Fatemeh, S. R., Saifullah, R., Abbas, F. M. A., & Azhar, M. E. (2012). Total phenolics, flavonoids and antioxidant activity of banana pulp and peel flours: Influence of variety and stage of ripeness. *International Food Research Journal, 19*, 1041–1046.

Fernandes, P. A. R., Moreira, S. A., Fidalgo, L. G., Santos, M. D., Queirós, R. P., Delgadillo, I., & Saraiva, J. A. (2015). Food Preservation under pressure (hyperbaric storage) as a possible improvement/alternative to refrigeration. *Food Engineering Reviews, 7*, 1–10.

Fernández-Falcón, M., Borges, A. A., & Borges-Pérez, A. (2003). Induced resistance to Fusarium wilt of banana by exogenous applications of indole acetic acid. *Phytoprotection, 84*, 149–153.

Ferris, R. S. B., Wainwright, H., & Thompson, A. K. (1993). Effect of maturity, damage and humidity on ripening of plantain and cooking banana. *ACIAR Proceedings, 50*, 434–437.

Ferris, R. S. B., Hotsouyame, G. K., Wainwright, H., & Thompson, A. K. (1993a). The effect of genotype, damage, maturity and environmental conditions on the postharvest life of plantain. *Tropical Agriculture, 70*, 45–50.

Ferris, R. S. B., Wainwright, H., & Thompson, A. K. (1995). The effects of morphology, maturity and genotype on the ripening and susceptibility of plantains (AAB) to mechanical damage. *Fruits, 50*, 45–50.

Finger, F. L., Puschmann, R., & Santos Barros, R. (1995). Effects of water loss on respiration, ethylene production and ripening of banana fruit. *Revista Brasileira de Fisiologia Vegetal, 7*, 115–118.

Forster, M., Rodríguez-Rodríguez, E. M., Darias-Martín, J., & Díaz, C. (2003). Distribution of nutrients in edible banana pulp. *Food Technology and Biotechnology, 41*, 167–171.

Frenkel, C. (1975). Oxidative turnover of auxins in relation to the onset of ripening in Bartlett pear. *Plant Physiology, 55*, 480–484.

Fuchs, Y., Zauberman, G., Yanko, U., & Homsky, S. (1975). Ripening of mango fruits with ethylene. *Tropical Science, 17*, 211–216.

Fukao, T., & Bailey-Serres, J. (2004). Plant responses to hypoxia – is survival a balancing act? *Trends in Plant Science, 9*, 449–456.

Gandhi, S., Sharma, M., & Bhatnagar, B. (2016). Comparative study on the ripening ability of banana by artificial ripening agent (calcium carbide) and natural ripening agents. *Indian Journal of Nutrition and Dietetics, 3*, 127.

Gane, R. (1934). Production of ethylene by some ripening fruits. *Nature, 134*, 1008.

Gane, R. (1936). A study of the respiration of bananas. *New Phytologist, 35*, 383–402.

García-Salinas, C., Ramos-Parra, P. A., & Díaz de la Garza, R. I. (2016). Ethylene treatment induces changes in folate profiles in climacteric fruit during postharvest ripening. *Postharvest Biology and Technology, 118*, 43–50.

Gautam, D. M., & Dhakal, D. D. (1993). *Falful tatha audhogik bali*. Bharatpur: Pavitra and Rupa Publication.

George, J. B. (1981). Storage and ripening of plantains. *University of London UK, PhD thesis*.

George, J. B., & Marriott, J. (1985). The effect of some storage conditions on the storage life of plantains. *Acta Horticulturae, 158*, 439–448.

Giovannoni, J. (2001). Molecular biology of fruit maturation and ripening. *Annual Review of Plant Physiology and Plant Molecular Biology, 52*, 725–749.

Golding, J. B., Shearer, D., Wyllie, S. G., & McGlasson, W. B. (1998). Application of 1-MCP and propylene to identify ethylene-dependent ripening processes in mature banana fruit. *Postharvest Biology and Technology, 14*, 87–98.

Golding, J. B., Shearer, D., McGlasson, W. B., & Wyllie, S. G. (1999). Relationships between respiration, ethylene, and aroma production in ripening banana. *Journal of Agriculture and Food Chemistry, 47*, 1646–1651.

Goonatilake, R. (2008). Effects of diluted ethylene glycol as a fruit-ripening agent. *Global Journal of Biotechnology & Biochemistry, 3*, 08–13.

Goswami, M. K., & Handique, F. J. (2013). Explants size response to *in vitro* propagation of *Musa* (AAA Group) 'Amritsagar' *Musa* (AAB Group) 'Malbhog' and *Musa* (AAB Group) 'Chenichampa' banana. *Indian Journal of Applied Research, 3*, 40–43.

Griffee, P. J., & Burden, O. J. (1976). Fungi associated with crown rot of boxed bananas in the Windward Islands. *Phytopathology Z, 85*, 149–158.

Griffee, P. J., & Pinegar, J. A. (1974). Fungicides for the control of the banana Crown Rot complex; *in vivo* and *in vitro* studies. *Tropical Science, 16*, 107–120.

Gross, J., & Flugel, M. (1982). Pigment changes in peel of the ripening banana (*Musa cavendish*). *Gartenbauwissenschaft, 47*, 62–64.

Gunasekara, S. R. W., Hemamali, K. K. G. U., Dayananada, T. G., & Jayamanne, V. S. (2015). Postharvest quality analysis of 'Embul' banana following artificial ripening techniques. *International Journal of Science, Environment and Technology, 4*, 1625–1632.

Hamilton, A. J., Lycett, G. W., & Grierson, D. (1990). Antisense gene that inhibits synthesis of the hormone ethylene in transgenic plants. *Nature, 346*, 284–287.

Han, Y.-c., Chang-chun, F., Kuang, J.-f., Chen, J.-y., & Lu, W.-j. (2016). Two banana fruit ripening-related C_2H_2 zinc finger proteins are transcriptional repressors of ethylene biosynthetic genes. *Postharvest Biology and Technology, 116*, 8–15.

Hardenburg, R. E., Watada, A. E., & Wang C. Y. (1990). *The commercial storage of fruits, vegetables and florist and nursery stocks*. United States Department of Agriculture, Agricultural Research Service, Agriculture Handbook 66.

Hardisson, A., Rubio, C., Baez, A., Martin, M., Alvarez, R., & Diaz, E. (2001). Mineral composition of the banana (*Musa acuminata*) from the island of Tenerife. *Food Chemistry, 73*, 153–161.

Harris, D. R., Seberry, J. A., Wills, L. J., & Spohr, L. J. (2000). Effect of fruit maturity on efficiency of 1-methylcyclopropene to delay the ripening of bananas. *Postharvest Biology and Technology, 20*, 303–308.

Harvey R. B. (1928). *Artificial ripening of fruits and vegetables*. https://conservancy.umn.edu/bitstream/handle/11299/184046/mn_1000_b_247.pdf?sequence=1 Accessed 18 August 2019

Heliofabia Virginia De Vasconcelos Facundo, Gurak, P. D., Mercadante, A. Z., Lajolo, F. M., & Cordenunsi, B. R. (2015). Storage at low temperature differentially affects the colour and carotenoid composition of two cultivars of banana. *Food Chemistry, 170*, 102–109.

Heslop-Harrison, J. S., & Schwarzacher, T. (2007). Domestication, genomics and the future for banana. *Annals of Botany, 100*, 1073–1084.

Hesselman, C. W., & Freebairn, H. T. (1969). Rate of ripening of initiated bananas as influenced by oxygen and ethylene. *Journal of the American Society for Horticultural Science, 94*, 635.

Ho, B. T., Joyce, D. C., & Bhandari, B. R. (2011). Release kinetics of ethylene gas from ethylene-alpha-cyclodextrin inclusion complexes. *Food Chemistry, 129*(2), 259–266.

Ho, B. T., Yuwono, T. D., Joyce, D. C., & Bhandari, B. R. (2015). Controlled release of ethylene gas from the ethylene-alpha-cyclodextrin inclusion complex powder with deliquescent salts. *Journal of Inclusion Phenomena and Macrocyclic Chemistry, 83*(3–4), 281–288.

Ho, B. T., Hofman, P. J., Joyce, D. C., & Bhandari, B. R. (2016). Uses of an innovative ethylene-α-cyclodextrin inclusion complex powder for ripening of mango fruit. *Postharvest Biology and Technology*, 113 77–86. Corrigendum to "Uses of an innovative ethylene-α-cyclodextrin inclusion complex powder for ripening of mango fruit". *Postharvest Biology and Technology, 117*, 239–239.

Hong, J. H., Hwanga, S. K., Chunga, G., & Cowan, A. K. (2007). Influence of lysophosphatidyl-ethanolamine application on fruit quality of Thompson seedless grapes. *Journal of Applied Horticulture, 9*, 112–114.

Hubbard, N. L., Pharr, D. M., & Huber, S. C. (1990). Role of sucrose phosphate synthase in sucrose biosynthesis in ripening bananas and its relationship to the respiratory climacteric. *Plant Physiology, 94*, 201–208.

Huber, D. J. (1983). Polyuronide degradation and hermicellulose modifications in ripening tomato fruits. *Journal of the American Society for Horticultural Science, 108*, 405–409.

Huber, O., & Mbéguié-A-Mbéguié, D. (2012). Expression patterns of ethylene biosynthesis genes from bananas during fruit ripening and in relationship with finger drop. *AOB Plants*. https://doi.org/10.1093/aobpla/pls041.

Hulme, A. C. (1971). *The biochemistry of fruits and their products* (Vol. 2). London: Academic.

Huseyin, P., Kurtoglu, S., Yagmur, F., Gumuş, H., Kumandaş, S. & Poyrazoglu, M. H. (2007). Calcium carbide poisoning via food in childhood. *Journal of Emergency Medicine, 32*, 145–149.

Imahori, Y., Yamamoto, K., Tanaka, H., & Bai, J. (2013). Residual effects of low oxygen storage of mature green fruit on ripening processes and ester biosynthesis during ripening in bananas. *Postharvest Biology and Technology, 77*, 19–27.

Imam, M. Z., & Akter, S. (2011). *Musa paradisiaca* L. and *Musa sapientum* L.: A phytochemical and pharmacological review. *Journal of Applied Pharmaceutical Science, 1*, 14–20.

Inaba, A., & Nakamura, R. (1986). Effect of exogenous ethylene concentration and fruit temperature on the minimum treatment time necessary to induce ripening in banana fruit. *Journal of the Japanese Society for Horticultural Science, 55*, 348–354.

Inaba, A., & Nakamura, R. (1988). Numerical expression for estimating the minimum ethylene exposure time necessary to induce ripening in banana fruit. *Journal of the American Society for Horticultural Science*, 561–564.

Iqbal, N., Khan, N., Ferrante, A., Trivellini, A., Francini, A., & Khan, M. I. (2017). Ethylene role in plant growth, development and senescence: Interaction with other phytohormone. *Frontiers in Plant Science, 8*, 475. https://doi.org/10.3389/fpls.2017.00475. Accessed 18 August 2019.

Islam, M. N., Rahman, A. H. M. S., Mursalat, M., Rony, A. H., & Khan, M. S. (2016). A legislative aspect of artificial fruit ripening in a developing country like Bangladesh. *Chemical Engineering Research Bulletin, 18*, 30–37.

Izonfuo, W. A. L., & Omuaru, V. O. T. (1988). Effect of ripening on the chemical composition of plantain peels and pulps (*Musa paradisiaca*). *Journal of the Science of Food and Agriculture, 45*, 333–336.

Jansasithorn, R., & Kanlavanarat, S. (2006). Effect of 1-MCP on physiological changes in banana ´Khai´. *Acta Horticulturae, 712*, 723–728.

Janssen, S., Tessmann, T., & Lang, W. (2014). High sensitive and selective ethylene measurement by using a large-capacity-on-chip preconcentrator device. *Sensors and Actuators B: Chemical, 197*, 405–413.

Jayanty, S., Song, J., Rubinstein, N. M., Chong, A., & Beaudry, R. M. (2002). Temporal relationship between ester biosynthesis and ripening events in bananas. *Journal of the American Society for Horticultural Science, 127*, 998–1005.

Jedermann, R., Geyer, M., Praeger, U., & Lang, W. (2013). Sea transport of bananas in containers – parameter identification for a temperature model. *Journal of Food Engineering, 115*, 330–338.

Jedermann, R., Dannies, A., Moehrke, A., Praeger, U., Geyer, M., & Lang, W. (2014). Supervision of transport and ripening of bananas by the Intelligent Container Supervision+of+transport+and + ripening+of+bananas+by+the+Intelligent+Container+ &pc = MOZD&form = MOZTSB. Accessed 24 March 2019.

Jedermann, R., Praeger, U., Geyer, M., Moehrke, A., & Lang, W. (2015). The intelligent container for banana transport supervision and ripening. *Acta Horticulturae, 1091*, 213–220.

Jiang, Y., Joyce, D. C., & Macnish, A. (1999a). Responses of banana fruit to treatment with 1-MCP. *Plant Growth Regulation, 28*, 77–82.

Jiang, Y., Joyce, D. C., & Macnish, A. (1999b). Extension of the shelf-life of banana fruit by 1-MCP in combination with polyethylene bags. *Postharvest Biology and Technology, 16*, 187–193.

Jiang, Y., Joyce, D. C., & Macnish, A. J. (2000). Effect of abscisic acid on banana fruit ripening in relation to the role of ethylene. *Journal of Plant Growth Regulation, 19*, 106–111.

Jiang, Y., Joyce, D. C., Jiang, W., & Lu, W. (2004a). Effects of chilling temperatures on ethylene binding by banana fruit. *Plant Growth Regulation, 43*, 109–115.

Jiang, Y., Joyce, D. C., Jiang, W., & Lu, W. (2004b). Effects of chilling temperatures on ethylene binding by banana fruit. *Plant Growth Regulation, 43*, 109–115.

Johns, G. G., & Scott, K. J. (1989). Delayed harvesting of bananas with sealed covers on bunches. 1. Modified atmosphere and microclimate inside sealed covers. *Australian Journal of Experimental Agriculture, 29*, 719–726.

Johnson, P., & Ecker, J. R. (1998). The ethylene gas signaling pathway in plants: A molecular perspective. *Annual Review of Genetics, 32*, 227–254.

Jones, D. R. (Ed.). (2000). *Diseases of banana, abacá and enset*. Wallingford: CABI Publishing.

Jones, D. R. (2009). Diseases and pests: Constraints to banana production. *Acta Horticulturae, 828*, 21–36.

Joomwong, A., & Joomwong, J. (2008). Physical, chemical properties and sensory analysis of banana [*Musa* (ABBB group) 'Kluai Teparod'] fruit. *Acta Horticulturae, 768*, 247–250.

Jordan, R., Seelye, R., & McGlone, A. (2001). A sensory-based alternative to brix/acid ratio. *Food Technology, 55*, 36–44.

Joyce, D. C., Macnish, A. J., Hofman, P. J., Simons, D. H., & Reid, M. S. (1999). Use of 1-methylcyclopropene to modulate banana ripening. In A. K. Kanellis, C. Chang, H. Klee, A. B. Bleecker, J. C. Pech, & D. Grierson (Eds.), *Biology and biotechnology of the plant hormone Ethylene II* (pp. 189–190). Dordrecht: Kluwer Academic Publishers.

Junior, A. V., Nascimento, J. R. O. D., & Lajolo, F. M. (2006). Molecular cloning and characterization of an α-amylase occurring in the pulp of ripening bananas and its expression in *Pichia pastoris. Journal of Agricultural and Food Chemistry, 54*, 8222–8228.

Kader, A. A. (1983). Physiological and biochmical effects of carbon monoxide added to controlled atmospheres of fruit. *Acta Horticulturae, 138*, 221–226.

Kader, A. A. (1987). Respiration and gas exchange in vegetables. In J. Weichmann (Ed.), *Postharvest physiology of vegetables* (pp. 25–44). New York: Marcel Dekker, Inc.

Kader, A. A. (Ed.). (1992). Postharvest biology and technology – an over-view. In *Post harvest technology of horticulture crops* (pp. 39–48). Division of Agriculture and Natural Resources, University of California.

Kahan, R. S., Nadel-Shifman, M., Temkin-Gorodeiski, N., Eisenberg, E., Zauberman, G., & Aharoni, Y. (1968). Effects of radiation on the ripening of bananas and avocado pears. In *Preservation of fruits and vegetables by irradiation* (pp. 3–11). Vienna: International Atomic Energy Agency.

Kamdee, C., Ketsa, S., & van Doorn, W. G. (2018). Effect of heat treatment on ripening and early peel spotting in Sucrier banana. *Postharvest Biology and Technology, 52*, 288–293.

Kanazawa, K., & Sakakibara, H. (2000). High content of dopamine, a strong antioxidant, in Cavendish banana. *Journal of Agricultural and Food Chemistry, 48*, 844–848.

Kanellis, A. K., Solomos, T., & Mattoo, A. K. (1989a). Hydrolytic enzyme activities and protein pattern of avocado fruit ripened in air and in low oxygen, with and without ethylene. *Plant Physiology, 90*, 259–266.

Kanellis, A. K., Solomos, T., Mehta, A. M., & Mattoo, A. K. (1989b). Decreased cellulase activity in avocado fruit subjected to 2.5 kPa O_2 correlates with lowered cellulase protein and gene transcripts levels. *Plant Cell Physiology, 30*, 817–823.

Kanellis, A. K., Loulakakis, K. A., Hassan, M., & Roubelakis-Angelakis, K. A. (1993). Biochemical and molecular aspects of low oxygen action on fruit ripening. In C. J. Pech, A. Latche, & C. Balague (Eds.), *Cellular and molecular aspects of the plant hormone ethylene* (pp. 117–122). Dordrecht: Kluwer Academic Publishers.

Kang, C. K., Yang, Y. L., Chung, G. H., & Palta, J. P. (2003). Ripening promoting and ethylene evolution in red pepper (*Capsicum annuum*) as influenced by newly developed formulations of a natural lipid, lysophosphatidylethanolamine. *Acta Horticulturae, 628*, 317–322.

Kao, H. Y. (1971). Extension of storage life of bananas by gamma irradiation. In *Disinfestation of fruit by irradiation* (pp. 125–136). Vienna: International Atomic Energy Agency.

Karikari, S. K., Marriot, J., & Hutchins, P. (1979). Changes during the respiratory climacteric in ripening plantain fruits. *Scientia Horticulturae, 10*, 369–376.

Karmawan, L. U., Suhandono, S., & Dwivany, F. M. (2009). Isolation of *MA-ACS* gene family and expression study of *MA-ACS1* gene in *Musa acuminata* cultivar Pisang Ambon Lumut. *HAYATI Journal of Biosciences, 16*, 35–39.

Kays, S. J., & Paull, R. E. (2004). Metabolic processes in harvested products. In *Postharvest biology* (pp. 79–136). Athens: Exon Press.

Ke, L. S., & Tsai, P. L. (1988). Changes in the ACC content and EFE activity in the peel and pulp of banana fruits during ripening in relation to ethylene production. *Journal of the Agricultural Association of China, 143*, 48–60.

Kendrick, M. D., & Chang, C. (2008). Ethylene signaling: New levels of complexity and regulation. *Current Opinion in Plant Biology, 11*, 479–485.

Kesar, R., Trivedi, P. K., & Nath, P. (2010). Gene expression of pathogenesis-related protein during banana ripening and after treatment with 1-MCP. *Postharvest Biology and Technology, 55*, 64–70.

Khan, M. Y., Khan, F. A., & Beg, M. S. (2013). Ethanol kerosene blends: Fuel option for kerosene wick stove. *International Journal of Engineering Research and Applications, 3*, 464–466.

Kirk-Othmer. (2004). *Encyclopedia of Chemical Technology* (Vol. 4, 5th ed.). New York: Wiley.

Kitinoja, L., & Kader, A. A. (2002). *Small-scale postharvest handling practices: A manual for horticultural crops* (4th Ed.). University of California, Davis Postharvest Technology Research and Information Center.

Klieber, A., Bagnato, N., Barrett, R., & Sedgley, M. (2002). Effect of post-ripening nitrogen atmosphere storage on banana shelf-life, visual appearance and aroma. *Postharvest Biology and Technology, 25*, 15–24.

Knee, M., Proctor, F. J., & Dover, C. J. (1985). The technology of ethylene control: use and removal in postharvest handling of horticultural commodities. *Annals of Applied Biology, 107*, 581–595.

Knight, C., Cutts, D. C., & Colhoun, J. (1977). The role of *Fusarium semitectum* in causing crown rot of bananas. *Phytopathologische Zeitschrift, 89*, 170–176.

Knowles, L., Trimble, M. R., & Knowles, N. R. (2001). Phosphorus status affects postharvest respiration, membrane permeability and lipid chemistry of European seedless cucumber fruit (*Cucumis sativus* L.). *Postharvest Biology and Technology, 21*, 179–188.

Kuang, J.-F., Chen, L., Shan, W., Yang, S., Lu, W.-j., & Chen, J.-y. (2013). Molecular characterization of two banana ethylene signaling component MaEBFs during fruit ripening. *Postharvest Biology and Technology, 85*, 94–101.

Kubo, Y., Akitsugu, I., & Nakamura, R. (1990). Respiration and C_2H_4 production in various harvested crops held in CO_2-enriched atmospheres. *Journal of the American Society for Horticultural Science, 115*, 975–978.

Kulkarni, S. G., Kudachikar, V. B., & Prakash, M. K. (2011). Studies on physico-chemical changes during artificial ripening of banana (*Musa* sp) variety 'Robusta'. *Journal of Food Science and Technology, 48*, 730–734.

Kumar, A., & Brahmachari, V. S. (2005). Effect of chemicals and packaging on ripening and storage behaviour of banana cv. Harichhaal (AAA) at ambient temperature. *Horticultural Journal, 18*, 86–90.

Lakade, A. J., Sundar, K., & Halady, P. (2018). Shetty Gold nanoparticle-based method for detection of calcium carbide in artificially ripened mangoes (Magnifera indica). *Food Additives & Contaminants, 35*, 1078–1084.

Larotonda, F. D. S., Genena, A. K., Dantela, D., Soares, H. M., Laurindo, J. B., Moreira, R. F. P. M., & Ferreira, S. R. S. (2008). Study of banana (Musa aaa Cavendish cv Nancia) trigger ripening for small scale process. *Brazilian Archives of Biology and Technology, 51*, 1033–1047.

Lee, P. J. (2008). Facts about banana potassium. http://ezinearticlescom/?Facts-About-Banana-Potassium&id=1762995 Accessed Oct 2009.

Leng, P., Yuan, B., & Guo, Y. (2014). The role of abscisic acid in fruit ripening and responses to abiotic stress. *Journal of Experimental Botany, 65*, 4577–4588.

Leong, L. P., & Shui, G. (2002). An investigation of antioxidant capacity of fruits in Singapore markets. *Food Chemistry, 76*, 69–75.

Letham, D. S. (1969). Influence of fertilizer treatment on apple fruit composition and physiology. II. Influence on respiration rate and contents of nitrogen, phosphorus and titratable acidity. *Australian Journal of Agricultural Research, 20*, 1073–1085.

Lin, W. C., & Ehret, D. L. (1991). Nutrient concentration and fruit thinning affect shelf-life of long English cucumber. *HortScience, 26*, 1299–1300.

Littmann, M. D. (1972). Effect of water loss on the ripening of climacteric fruits. *Queensland Journal of Agriculture and Animal Science, 29*, 103–113.

Liu, F. W. (1976a). Correlation between banana storage life and minimum treatment time required for ethylene response. *Journal of the American Society for Horticultural Science, 101*, 63–65.

Liu, F. W. (1976b). Banana response to low concentration of ethylene. *Journal of the American Society for Horticultural Science, 101*, 222–225.

Liu, F. W. (1976c). Storing ethylene pretreated bananas in controlled atmosphere and hypobared air. *Journal of the American Society for Horticultural Science, 101*, 198–201.

Liu, T.-T., & Yang, T.-S. (2002). Optimization of solid-phase microextraction analysis for studying change of headspace flavor compounds of banana during ripening. *Journal of Agricultural and Food Chemistry, 50*, 653–657.

Liu, X., Shiomi, S., Nakatsuka, A., Kubo, Y., Nakamura, R., & Inaba, A. (1999). Characterization of ethylene biosynthesis associated with ripening in banana fruit. *Plant Physiology, 121*, 1257–1265.

Liu, R., Wang, Y., Qin, G., & Tian, S. (2016). Molecular basis of 1-methylcyclopropene regulating organic acid metabolism in apple fruit during storage. *Postharvest Biology and Technology, 117*, 57–63.

Lizada, M. C. C., Pantastico, E. B., Abdullah Shukor, A. R., & Sabari, S. D. (1990). Ripening of banana. In H. Abdulla & E. B. Pantastico (Eds.), *Banana* (pp. 65–84). Association of Southeast Asian Nations Food Handling Bureau.

Lohani, S., Trivedi, P. K., & Nath, P. (2004). Changes in activities of cell wall hydrolases during ethylene-induced ripening in banana: Effect of 1-MCP, ABA and IAA. *Postharvest Biology and Technology, 31*, 119–126.

Lòpez-Gòmez, R., Campbell, A., Dong, J. G., Yang, S. F., & Gòmez-Lim, M. A. (1997). Ethylene biosynthesis in banana fruit: Isolation of a genomic clone to ACC oxidase and expression studies. *Plant Science, 123*, 123–131.

Loulakakis, C. A., Hassan, M., Gerasopoulos, D., & Kanellis, A. K. (2006). Effects of low oxygen on in vitro translation products of poly(A) + RNA, cellulase and alcohol dehydrogenase expression in preclimacteric and ripening-initiated avocado fruit. *Postharvest Biology and Technology, 39*, 29–37.

Lu and Jian-ye Chen. (2013). Molecular characterization of two banana ethylene signaling component MaEBFs during fruit ripening. *Postharvest Biology and Technology, 85*, 94–101.

Lurie, S. (2008). Regulation of ethylene biosynthesis in fruits by aminoethoxyvinyl glycine and 1-Methylcyclopropene. *Acta Horticulturae, 796*, 31–41.

Macklin, G. (2001). SmartAir offers in-transit banana ripening. https://www.refrigeratedtransporter.com/archive/smartair-offers-transit-banana-ripening. Accessed 24 March 2019.

Madamba, S. P., Baes, A. U., & Mendoza, D. B., Jr. (1977). Effect of maturity on some biochemical changes during ripening of banana *Musa sepientum* cv. Lakatan. *Food Chemistry, 2*, 177–183.

Maersk. (2019). The perfect environment for your bananas. https://www.maersk.com/en/solutions/shipping/ocean-transport/refrigerated-cargo/bananas-and-pineapples. Accessed 24 March 2019.

Maia, V. M., Salomão, L. C. C., Siqueira, D. L., Aspiazúl, I., & Maia, L. C. B. (2014). Physical and metabolic changes induced by mechanical damage in 'Dwarf-Prata' banana fruits kept under cold storage. *Australian Journal of Crop Science, 8*, 1029–1037.

Maneenuam, T., & Doorn, S. K. (2007). High oxygen levels promote peel spotting in banana fruit. *Postharvest Biology and Technology, 43*, 128–132.

Maneenuam, T., Ketsa, S., & Van Doorn, W. G. (2007). High oxygen levels promote peel spotting in banana fruit. *Postharvest Biology and Technology, 43*, 128–132.

Manjunatha, G., Lokesh, V., & Bhagyalakshmi, N. (2012). Nitric oxide-induced enhancement of banana fruit attributes and keeping quality. *Acta Horticulturae, 934*, 799–806.

Manrique-Trujillo, S. M., Ramírez-López, A. C., Ibarra-Laclette, E., & GómezLim, M. A. (2007). Identification of genes differentially expressed during ripening of banana. *Journal of Plant Physiology, 164*, 1037–1050.

Maqbool, M., Ali, A., Ramachandran, S., Smith, D. R., & Alderson, P. G. (2010). Control of post-harvest anthracnose of banana using a new edible composition coating. *Crop Protection, 29*, 1136–1141.

Marchal, J., & Mallessard, R. (1979). Comparison des immobilisations minerales de quatre cultivars de bananiers a fruits pour cuisson et de deux 'Cavendish'. *Fruits, 34*, 373–392.

Marchal, J., & Nolin, J. (1990). Fruit quality. Pre- and post-harvest physiology. *Fruits* Special issue, 119–122.

Marchal, J., Nolin, J., & Letorey, J. (1988). Influence sur la maturation de l'enrobage de bananes avec du Semperfresh. *Fruits, 43*, 447–453.

Marcus, Y. (1990). Recommended methods for the purification of solvents and tests for impurities: 1, 2ethanediol and 2, 2, 2-trifluoroethanol. *Pure and Applied Chemistry, 62*, 139–147.

Marriott, J., & New, S. (1975). Storage physiology of bananas from new tetrapolid clones. *Tropical Science, 17*, 155–163.

Marriott, J., New, S., Dixon, E. A., & Martin, K. J. (1979). Factors affecting the preclimacteric period of banana fruit bunches. *Annals of Applied Biology, 93*, 91–100.

Mathooko, F. M., Tsunashima, Y., Kubo, Y., & Inaba, A. (2004). Expression of a 1-aminocyclop ropane-1-carboxylate (ACC) oxidase gene in peach (*Prunus persica* L.) fruit in response to treatment with carbon dioxide and 1- methylcyclopropene: possible role of ethylene. *African Journal of Biotechnology, 3*, 497–502.

Matsumura, S., Tomizawa, N., Toki, A., Nishikawa, K., & Toshima, K. (1999). Novel poly(vinyl alcohol)-degrading enzyme and the degradation mechanism. *Macromolecules, 23*, 7753–7761.

Maxie, E. C., & Sommer, N. F. (1968). Changes in some chemical constituents in irradiated fruits and vegetables. In *Preservation of fruits and vegetables by radiation* (pp. 39–56). Vienna: International Atomic Energy Agency.

Maxie, E. C., Amezquita, R., Hassan, B. M., & Johnson, C. F. (1968). Effect of gamma irradiation on the ripening of banana fruits. *Proceedings of the American Society for Horticultural Science, 92*, 235–244.

Mbéguié-A-Mbéguié, D., Olivier Hubert, O., Fils-Lycaon, B., Chillet, M., & Baurens, F. C. (2008). EIN3-likegene expression during fruit ripening of Cavendish banana (*Musa acuminata* cv. Grande naine). *Physiologia Plantarum, 133*, 435–448.

Mbéguié-A-Mbéguié, D., Hubert, O., Baurens, F. C., Sidibé-Bocs, S., Matsumoto, T., Chillet, M., & Fils-Lycaon, B. (2009). Expression patterns of cell wall modifying genes from banana during ripening in relationship with finger drop. *Journal of Experimental Botany, 60*, 2021–2034.

McCarthy, A. L., Palmer, J. K., Shaw, C. P., & Anderson, E. E. (1963). Correlation of gas chromatographic data with flavour profiles of fresh banana fruit. *Journal of Food Science, 28*, 379–384.

McMurchie, E. J., McGlasson, W. B., & Eaks, I. L. (1972). Treatment of fruit with propylene gives information about the biogenesis of ethylene. *Nature, 237*, 235–236.

Medlicott, A. P., Sigrist, J. M. M., Reynolds, S. B., & Thompson, A. K. (1987). Effects of ethylene and acetylene on mango fruit ripening. *Annals of Applied Biology, 111*, 439–444.

Medlicott, A. P., Semple, A. J., Thompson, A. J., Blackbourne, H. R., & Thompson, A. K. (1992). Measurement of colour changes in ripening bananas and mangoes by instrumental, chemical and visual assessments. *Tropical Agriculture, 69*, 161–166.

Mercantilia. (1989). *Guide to food transport – fruit and vegetables*. Copenhagen: Mercantilia Publishers.

Meredith, D. S. (1963). Latent infections in *Pyricularia grisea* causing pitting disease of banana fruits in Costa Rica. *Plant Disease Reporter, 47*, 766–768.

Minas, I. S., Font, I., Forcada, C., Dangl, G. S., Gradziel, T. M., Dandekar, A. M., & Crisosto, C. H. (2015). Discovery of non-climacteric and suppressed climacteric bud sport mutations originating from a climacteric Japanese plum cultivar (*Prunus salicina* Lindl.) Frontiers of Plant Science 6, 316 https://doi.org/10.3389/fpls.2015.00316 Accessed 4 May 2019.

Mitchell, F. G. (1990). Postharvest physiology and technology of kiwifruit. *Acta Horticulturae, 282*, 291–307.

Mladenoska, I. (2013). The preservation of the whole banana fruits by utilization of coconut oil-beeswax edible coatings. Proceedings of the 10th Symposium "Novel technologies and Economic Development", Leskovac, pp. 13–20.

Moehrke, A. (2014). *Process for ripening bananas inside of a shipping container*. U.S. Provisional Patent Application No. 61/801,515, filed March 15, 2013, https://patents.google.com/patent/WO2014144788A1/en. Accessed 23 March 2019.

Molina, G. (2019). Bioversity International – Asia Pacific, Philippine Office: http://banana-networks.org/Bapnet/2016/02/05/kluai-namwa-foc-tr4-resistant-variety/. Accessed 11 March 2019.

Montenegro, E. H. (1988). *Postharvest behaviour of banana harvested at different stages of maturity*. BS student project, University of the Phillipines, Los Baños, Laguna.

Moradinezhad, F., Sedgley, M., Klieber, A., & Able, A. J. (2008). Variability of responses to 1-methycyclopropene by banana: Influence of time of year at harvest and fruit position in the bunch. *Annual Applied Biology, 152*, 223–234.

Moric, C. L. S., dos Passosa, N. A., Oliveirab, J. E., Mattosod, L. H. C., Moric, F. A., Carvalhoc, A. G., Fonsecac, A. S., & Tonolica, G. H. D. (2014). Electrospinning of Zzein/tannin bio-nanofibers. *Industrial Crops and Products, 52*, 298–304.

Morris, L., Yang, S. F., & Mansfield, D. (1981). Postharvest physiology studies. Californian fresh market tomato advisory board annual report 1980–1981, 85–105.

Morton, J. F. (1987). *Fruits of warm climates*. http://www.hort.purdue.edu/newcrop/morton/papaya_ars.html. Accessed 29 March 2019.

Mura, K., & Tanimura, W. (2003). Changes in polyphenol compounds in banana pulp during ripening. *Food Preservation Science, 29*, 347–351.

Murata, T. (2006). Physiological and biochemical studies of chilling injury in bananas. *Physiologia Plantarum, 22*, 401–411.

Mustaffa, R., Osman, A., Yusof, S., & Mohamed, S. (1998). Physico-chemical changes in Cavendish banana (*Musa cavendishii* L var. Montel) at different positions within a bunch during development and maturation. *Journal of the Science of Food and Agriculture, 78*, 201–207.

Nair, H., & Tung, H. F. (1988). Postharvest physiology and storage of Pisang Mas. Proceedings of the UKM simposium Biologi Kebangsaan ketiga, Kuala Lumpur, Nov. 22–24.

Narasimham, P., Dalal, V. B., Nagaraja, N., Krishnaprakash, M. S., & Amla, B. L. (1971). *Effects of smoking on some physiological changes in bananas*. *Journal of Food Science and Technology, 8*, 84–85.

Ncama, K., Magwaza, L. S., Mditshwa, A., & Tesfay, S. Z. (2018). Plant-based edible coatings for managing postharvest quality of fresh horticultural produce. *Food Packaging and Shelf-life, 16*, 157–167.

Neljubow, D. (1901). Ueber die horizontale nutation der stengel von *Pisum sativum* und einiger Anderer. *Pflanzen Beih Bot Zentralbl, 10*, 128–139.

Nelson, S. (n.d.). Banana ripening: Principles and practice. University of Hawaii file:///C:/Users/Anthony%20Keith%20Thomps/Downloads/Banana%20ripeningbunchmanagement.pdf. Accessed 20 Feb 2019.

Nelson, S. C., & Javier, F. B. (2007). Trials of FHIA banana varieties for resistance to black leaf streak in Pohnpei FSM. 38th Hawaii Banana Industry Association Annual Conference, Kalaeloa, Hawaii, Aug. 24, 2007. https://www.ctahr.hawaii.edu/nelsons/HBIA_2007_Nelson_Javier.pdf. Accessed 28 March 2019.

New, S., & Marriott, J. (1974). Post-harvest physiology of tetraploid banana fruit: Response to storage and ripening. *Annals of Applied Biology, 78*, 193–204.

New, S., & Marriott, J. (1983). Factors affecting the development of finger drop in bananas after ripening. *Journal of Food Technology, 18*, 241–250.

Nham, N. T., Willits, N., Zakharov, F., & Mitcham, E. J. (2017). A model to predict ripening capacity of 'Bartlett' pears (*Pyrus communis* L.) based on relative expression of genes associated with the ethylene pathway. *Postharvest Biology and Technology, 128*, 138–143.

Nogueira, J. M., Fernandes, P. J., & Nascimento, A. M. (2003). Composition of volatiles of banana cultivars from Madeira Island. *Phytochemical Analysis, 14*, 87–90.

Nolin, J. (1985). Etat de maturite des bananes (Giant Cavendish) a la recolte, une nouvelle methode de mesure. *Fruits, 40*, 623–631.

Noysang, C., Buranasukhon, W., & Khuanekkaphan, M. (2019). Phytochemicals and pharmacological activities from banana fruits of several *Musa* species for using as cosmetic raw materials. *Applied Mechanics and Materials, 891*, 30–40.

Nura, A., Dandag, M. A., & Wali, N. R. (2018). Effects of artificial ripening of banana (*Musa* spp) using calcium carbide on acceptability and nutritional quality. *Journal of Postharvest Technology, 6*, 14–20.

Nyanjage, M. O., Wainwright, H., Bishop, C. F. H., & Cullum, F. J. (2000). A comparative study on the ripening and mineral content of organically and conventionally grown Cavendish bananas. *Biological Agriculture and Horticulture, 18*, 221–234.

Obando, J., Fernández-Trujillo, J. P., Martínez, J. A., Alarcón, A. L., Eduardo, I., Arús, P., & Monforte, A. J. (2008). Identification of melon fruit quality quantitative trait loci using near-isogenic lines. *Journal of the American Society for Horticultural Science, 133*, 139–151.

Ozgen, M., Farag, K. M., Ozgen, G., & Palta, J. P. (2005). Lysophosphatidylethanolamine accelerates color development and promotes shelf life of cranberries. *HortSciences, 40*, 127–130.

Palmer, J. K. (1971). The banana. In A. C. Hulme (Ed.), *The biochemistry of fruits and their products* (Vol. 2). Academic: London.

Palomer, X., Roig-Villanova, I., Grima-Calvo, D., & Vendrell, M. (2005). Effects of nitrous oxide treatment on the postharvest ripening of banana fruit. *Postharvest Biology and Technology, 36*, 167–175.

Pan, S. L., Huang, C. Y., Wang, H. B., Pang, X. Q., Huang, X. M., & Zhang, Z. Q. (2007). Hydrogen peroxide induced chilling-resistance of postharvest banana fruit. *Journal of South China Agricultural University, 28*, 34–37.

Pantastico, E. B. (1975). Editor – *Postharvest physiology, handling and utilization of tropical and sub-tropical fruits and vegetables*. Westpoint: AVI Publishing Co.

Parmar, B. R., & Chundawat, B. S. (1984). Effect of growth regulators and sleeving on maturity and quality of banana cv. Basrai. *South Indian Horticulture, 32*, 201–204.

Pasentsis, K., Falara, V., Pateraki, I., Gerasopoulos, D., & Kanellis, A. K. (2007). Identification and expression profiling of low oxygen regulated genes from *Citrus* flavedo tissues using RT PCR differential display. *Journal of Experimental Botany, 58*, 2203–2216.

Patel, B. B., Ahlawat, T. R., & Patel, B. B. (2017). Effect of potash fertilizers and magnesium on quality of banana (*Musa paradisiaca* L.) Grand Naine. *International Journal of Chemical Studies, 5*, 1586–1591.

Pathak, N., Asif, M. H., Dhawan, P., Srivastava, M. K., & Nath, P. (2003). Expression and activities of ethylene biosynthesis enzymes during ripening in banana fruit and effect of 1-MCP treatment. *Plant Growth Regulation, 40*, 11–19.

Paul, V., Pandey, R., & Srivastava, G. C. (2012). The fading distinctions between classical patterns of ripening in climacteric and non-climacteric fruit and the ubiquity of ethylene-An overview. *Journal of Food Science and Technology, 49*, 1–21.

Paull, R. E. (1996). Ethylene, storage and ripening temperatures affect Dwarf Brazilian banana finger drop. *Postharvest Biology and Technology, 8*, 65–74.

Paull, R. E., & Chen, N. J. (2000). Heat treatment and fruit ripening. *Postharvest Biology and Technology, 21*, 21–37.

Peacock, B. C. (1972). Role of ethylene in the initiation of fruit ripening. *Queensland Journal of Agriculture and Animal Science, 29*, 137–145.

Pegoraro, C., dos Santos, R. S., Krüge, M. M., Tiecher, A., da Mai, L. C., Rombaldi, C. V., & de Oliveira, A. C. (2012). Effects of hypoxia storage on gene transcript accumulation during tomato fruit ripening. *Brazilian Journal of Plant Physiology, 24*(2). Campos dos Goytacazes Apr./June 2012. On-line version ISSN 1677–9452). https://doi.org/10.1590/S1677-04202012000200007.

Pelayo, C., Eduardo, V.de B. Vilas-Boas, Benichou, M., & Kader, A. A. (2003). Variability in responses of partially ripe bananas to 1-methylcyclopropene. *Postharvest Biology and Technology, 28*, 75–85.

Pereira, L. C., Ngoh Newilah, G. B., Davey, M. W., & Van den Bergh, I. (2011). Validation of rapid (colour-based) prescreening techniques for analysis of fruit provitamin A contents in banana (*Musa* spp.). *Acta Horticulturae, 897*, 161–168.

Peroni-Okita, F. H. G., Cardoso, M. B., Agopian, R. G. D., Louro, R. P., Nascimento, J. R. O., & Purgatto, E. (2013). The cold storage of green bananas affects the starch degradation during ripening at higher temperature. *Carbohydrate Polymers, 96*, 137–147.

Pinheiro, A. C. M., Boas, E.V.deB.V, & Mesquita, C. T. (2005). Action of 1-methylcyclopropene (1-MCP) on shelf life of 'Apple' banana. *Revista Brasileira de Fruticultura, 27*, 25–28.

Ploetz, R. C., Kepler, A. K., Daniels, J., & Nelson, S. C. (2007). *Banana and plantain – An overview with emphasis of Pacific Island*. Hōlualoa: Permanent Agriculture Resources.

Pokhrel, P. (2013). Use of higher ethylene generating fruits for ripening as an alternative to ethylene. *Journal of Food Science and Technology Nepal, 8*, 84–86.

Ponce-Valadez, M., Fellman, S. M., Giovannonic, J., Gana, S.-S., & Watkins, C. B. (2009). Differential fruit gene expression in two strawberry cultivars in response to elevated CO_2 during storage revealed by a heterologous fruit microarray approach. *Postharvest Biology and Technology, 51*, 131–140.

Pontes, M., Pereira, J., & Câmara, J. S. (2012). Dynamic headspace solid-phase microextraction combined with one-dimensional gas chromatography–mass spectrometry as a powerful tool to differentiate banana cultivars based on their volatile metabolite profile. *Food Chemistry, 134*, 2509–2520.

Porcher, M. H. (1998). The University of Melbourne. http://www.plantnames.unimelb.edu.au/Sorting/Musa-cvs.html. Accessed 17 April 2019.

Puraikalan, Y. (2018). Characterization of proximate, phytochemical and antioxidant analysis of banana (*Musa sapientum*) peels/skins and objective evaluation of ready to eat /cook product made with banana peels. *Current Research in Nutrition and Food Science, 6*. https://doi.org/10.12944/CRNFSJ.6.2.13.

Purgatto, E., Lajolo, F. M., Oliveira do Nascimento, J. R., & Cordenunsi, B. R. (2001). Inhibition of β-amylase activity, starch degradation and sucrose formation by indole-3-acetic acid during banana ripening. *Planta, 212*, 823–828.

Purgatto, E., Olivera do Nascimento, J. R., Lajolo, F. M., & Cordenunsi, B. R. (2002). The onset of starch degradation during banana ripening is concomitant to changes in the control of free and conjugated form of indole-3-acetic acid. *Journal of Plant Physiology, 159*, 1105–1111.

Quazi, M. H., & Freebairn, H. T. (1970). The influence of ethylene oxygen and carbon dioxide on ripening of bananas. *Botanical Gazette, 131*, 5–14.

Ram, H. B., Singh, S. K., Singh, R. V., & Surjeet, S. (1979). Effect of Ethrel and smoking treatment on ripening and storage of banana cultivar Himachal. *Progressive Horticulture, 11*, 69–75.

Ram, B. K. C., Gautam, D. M., & Tiwari, S. (2009). Use of Ethephone and indigenous plant materials in ripening banana in winter. *Nepal Agricultural Research Journal, 9*, 113.

Ramesh Kumar, A., & Kumar, N. (2007). Sulfate of potash foliar spray effects on yield, quality, and post-harvest life of banana. *Better Crops, 91*, 22–24.

Ramsey, M. D., Daniells, J. W., & Anderson, V. J. (1990). Effects of Sigatoka leaf spot (*Mycosphaerella musicola* Leach) on fruit yields, field ripening and greenlife of bananas in North Queensland. *Scientia Horticulturae, 41*, 305–313.

Reid, M. S. (2002). Ethylene in postharvest technology In: Kader, A.A. (ed). *Postharvest technology of horticultural crops*. University of California, Division of Agriculture and Natural Resources, Publication 3311. pp.149–162..

Renoo, Y., Kongpensook, V., Chambers IV, E., Suwansichon, S., Oupadissakoon, C & Retiveau, A. (n.d.). The sensory characteristics of commercially available bananas in Thailand. file:///C:/Users/Anthony%20Keith%20Thomps/Downloads/SSP2008-P05-Yenket.pdf. Accessed 18 April 2019.

Robinson, J. C., & Saúco, V. G. (2010). *Bananas and plantains* (2nd ed.). Wallingford: CAB International.

Romero, I., Sanchez-Ballesta, M. T., Maldonado, R., Escribano, M. I., & Merodio, C. (2008). Anthocyanin, antioxidant activity and stress-induced gene expression in high CO_2-treated table grapes stored at low temperature. *Journal of Plant Physiology, 165*, 522–530.

Romphophak, T., Siriphanich, J., Promdang, S., & Ueda, Y. (2004). Effect of modified atmosphere storage on the shelf life of banana 'Sucrier'. *Journal of Horticultural Science & Biotechnology, 79*, 659–663.

Rossetto, M. R. M., Purgatto, E., do Nascimento, J. R. O., Lajolo, F. M., & Cordenuns, B. R. (2003). Effects of gibberellic acid on sucrose accumulation and sucrose biosynthesizing enzymes activity during banana ripening. *Plant Growth Regulation, 41*, 207–214.

Rothan, C., Duret, S., Chevalier, C., & Raymond, P. (1997). Suppression of ripening associated gene expression in tomato fruit subjected to a high CO_2 concentration. *Plant Physiology, 114*, 255–263.

Salmon, B., Martin, G. J., Remaud, G., & Fourel, F. (1996). Compositional and isotopic studies of fruit flavours. Part I. The banana aroma. *Flavour and Fragrance Journal, 11*, 353–359.

Saltveit, M. E. (1999). Effect of ethylene on quality of fresh fruits and vegetables. *Postharvest Biology and Technology, 15*, 279–292.

Saltveit, M., Bradford, K. J., & Dilley, D. R. (1978). Silver ion inhibits ethylene synthesis and action in ripening fruits. *Journal of the American Society for Horticultural Science, 103*, 472–475.

Sanaeifar, A., Mohtasebi, S. S., Ghasemi-Varnamkhasti, M., Ahmadi, H., & Lozano, J. (2014). Development and application of a new low-cost electronic nose for the ripeness monitoring of banana using computational techniques (PCA, LDA, SIMCA, and SVM). *Czech Journal of Food Sciences, 32*, 538–548.

Sanchez-Nieva, F., Hernandez, L., & Bueso de Vinas, C. (1970). Studies on the ripening of plantains under controlled conditions. *Journal of Agriculture of the University of Puerto Rico, 54*, 517–529.

Saraiva, J. A. (2014). Storage of foods under mild pressure (hyperbaric storage) at variable (uncontrolled) room: a possible new preservation concept and an alternative to refrigeration? *Journal of Food Processing & Technology, 5*, 56. https://doi.org/10.4172/2157-7110.S1.002. (Abstract).

Sarananda, K. H. (1990). Effect of calcium carbide on ripening of 'Embul' bananas. *Tropical Agriculturist, 146*, 27–35.

Satyan, S., Scott, K. J., & Graham, D. (1992). Storage of banana bunches in sealed polyethylene tubes. *Journal of Horticultural Science, 67*, 283–287.

Scott, K. J., Blake, J. R., Strachan, G., Tugwell, B. L., & McGlasson, W. B. (1971). Transport of bananas at ambient temperatures using polyethylene bags. *Tropical Agriculture, 48*, 245–253.

Scriven, F. M., Gek, C. O., & Wills, R. B. H. (1989). Sensory differences between bananas ripened with and without ethylene. *HortScience, 24*, 983–984.

SeaLand. (1991). *Shipping guide to perishables*. SeaLand Services Inc., P.O. Box 800, Iselim, New Jersey 08830.

Seenappa, M., Laswai, M., & Fernando, S. P. F. (1986). Availability of L-ascorbic acid in Tanzanian banana. *Journal of Food Science and Technology, 23*, 293–295.

Segall, Y., Grendell, R. L., Toia, R. F., & Casida, J. E. (1991). Composition of technical ethephon [(2-chloroethyl) phosphonic acid] and some analogues relative to their reactivity and biological activity. *Journal of Agricultural and Food Chemistry, 39*, 380–385.

Semple, A. J., & Thompson, A. K. (1988). Influence of the ripening environment on the development of finger drop in bananas. *Journal of the Science of Food Agriculture, 46*, 139–146.

Senna, M. M. H., Al-Shamrani, K. M., & Al-Arifi, A. S. (2014). Edible coating for shelf-life extension of fresh banana fruit based on gamma irradiated plasticized poly(vinyl alcohol)/carboxymethyl cellulose/tannin composites. *Materials Sciences and Applications, 5*, 395–415.

Seo, Y., & Hosokawa, A. (1983a). Relationship between evolved CO_2 and increase in sugar content during artificial banana ripening. *Journal of the Japanese Society of Agricultural Machinery and Food Engineers, 44*, 633–638.

Seo, Y., & Hosokawa, A. (1983b). Prediction of sugar content with temperature schedule in artificial banana ripening. *Journal of the Japanese Society of Agricultural Machinery and Food Engineers, 45*, 229–234.

Seo, Y., Amano, T., Kawagoe, Y. & Sagara, Y. (1995). Artificial banana ripening by fuzzy control. Control applications in post-harvest and processing technology. 1st ifac/cigr/eurageng/ishs workshop, Ostend, 1–2 June 1995 J. de Baerdemaeker and J. Vandewalle (editors) 151–155.

Seymour, G. B. (1986). The effect of gases and temperature on banana ripening. *PhD thesis, University of Reading.*

Seymour, G. B., Thompson, A. K., & John, P. (1987). Inhibition of degreening in the peel of bananas ripened at tropical temperatures. 1. Effect of high temperature on changes in the pulp and peel during ripening. *Annals of Applied Biology, 110*, 145–151.

Seymour, G. B., John, P., & Thompson, A. K. (1987a). Inhibition of degreening in the peel of bananas ripened at tropical temperature. 2. Role of ethylene, oxygen and carbon dioxide. *Annals of Applied Biology, 110*, 153–161.

Seymour, G. B., Ryder, C. D., Cevik, V., Hammond, J. P., Popovich, A., King, G. J., Vrebalov, J., Giovannoni, J. J., & Manning, K. (2011). A SEPALLA total acidity gene is involved in the development and ripening of strawberry (*Fragaria×ananassa* Duch.) fruit, a non-climacteric tissue. *Journal of Experimental Botany, 62*, 1179–1188.

Seymour, G. B., Chapman, N. H., Chew, B. L., & Rose, J. K. C. (2013). Regulation of ripening and opportunities for control in tomato and other fruits. *Plant Biotechnology Journal, 11*, 269–278.

Sfakiotakis, E., Antunes, M. D., Stavroulakis, G., & Niklis, N. (1999). Rapporti fra produzione di etilene e maturazione dei frutti della cultivar Hayward nelle fasi di raccolta e conservazione. *Rivista di Frutticoltura e di Ortofloricoltura, 61*, 59–65.

Shaoying, Z., Zhu, L., & Dong, X. (2015). Combined treatment of carbon monoxide and chitosan reduced peach fruit browning and softening during cold storage. *International Journal of Nutrition and Food Sciences, 4*, 477–482.

Sheng, K., Zheng, H., Shui, S. S., Yan, L., & Zheng, L. (2018). Comparison of postharvest UV-B and UV-C treatments on table grape: changes in phenolic compounds and their transcription of biosynthetic genes during storage. *Postharvest Biology and Technology, 138*, 74–81.

Shiota, H. (1993). New esteric compounds in the volatiles of banana fruit (*Musa sapientum* L.). *Journal of Agricultural and Food Chemistry, 41*, 2056–2062.

Shorter, A. J., Scott, K. J., & Graham, D. (1987). Controlled atmosphere storage of bananas in bunches at ambient temperatures. *CSIRO Food Research Queensland, 47*, 61–63.

Siddiqui, M. W., & Dhua, R. S. (2010). Eating artificial ripened fruits is harmful. *Current Science, 99*, 1664–1668.

Sidhu, S. S., & Zafar, T. A. (2018). Bioactive compounds in banana fruits and their health benefits. *Food Quality and Safety, 2*, 183–188.

Silvis, H., & Thompson, A. K. (1974). Preliminary investigations on improving the quality of bananas for the local market in the Sudan. *Sudan Journal of Food Science and Technology, 6*, 7–17.

Simmonds, N. W. (1966). *Bananas* (2nd ed.). London: Longmans.

Simmonds, N. W., & Shepherd, K. (1955). The taxonomy and origins of the cultivated *bananas*. *Journal of the Linnaean Society, 55*, 302–312.

Sinclair, J. (1988). *Refrigerated transportation*. London: Container Marketing Limited.

Singh, B., Singh, J. P., Kaur, A., & Singh, N. (2016). Bioactive compounds in banana and their associated health benefits – A review. *Food Chemistry, 206*, 1–11.

Sisler, E. C. (1991). Ethylene-binding components in plants. In A. K. Mattoo & J. E. Suttle (Eds.), *The plant hormone ethylene* (pp. 81–99). Boca Raton: CRC Press.

Sisler, E. C., & Blankenship, S. M. (1993). Diazocyclopentadiene (DACP) a light sensitive reagent for the ethylene receptor in plants. *Plant Growth Regulation, 12*, 125–132.

Sisler, E. C., & Lallu, N. (1994). Effect of diazocyclopentadiene (DACP) on tomato fruits harvested at different ripening stages. *Postharvest Biology and Technology, 4*, 245–254.

Slaughter, M. L. D. C., & Thompson, J. F. (1997). Optical chlorophyll sensing system for banana ripening. *Postharvest Biology and Technology, 12*, 273–283.

Smith, N. J. S. (1989). Textural and biochemical changes during ripening of bananas. *University of Nottingham, PhD thesis*.

Smith, N. J. S., & Seymour, G. B. (1990). Cell wall changes in bananas and plantains. *Acta Horticulturae, 269*, 283–289.

Smith, N. J. S., & Thompson, A. K. (1987). The effects of temperature, concentration and exposure time to acetylene on initiation of banana ripening. *Journal of the Science of Food Agriculture, 40*, 43–50.

Smith, N. J. S., Seymour, G. B., & Thompson, A. K. (1986). Effects of high temperatures on ripening responses of bananas to acetylene. *Annals of Applied Biology, 108*, 667–672.

Snowdon, A. L. (1990). A colour atlas of postharvest diseases and disorders of fruits and vegetables. In *General introduction and fruits* (Vol. 1). London: Wolfe Scientific Ltd.

Sogo-Temi, C. M., Idowu, O. A., & Idowu, E. (2014). Effect of biological and chemical ripening agents on the nutritional and metal composition of banana (*Musa spp.*). *Journal of Applied Sciences and Environmental Management, 18*, 243–246.

Solomos, T., & Biale, J. B. (1975). Facteurs et regulation de la maturation des fruits. *Colloque Internationaux du Center National de la Recherche Scientifique, 238*, 221–228.

Soltani, M., Alimardani, R., & Omid, M. (2011). Evaluating banana ripening status from measuring dielectric properties. *Journal of Food Engineering, 105*, 625–631.

Song, M., Tang, L., Zhang, X., Bai, M., Pang, X., & Zhang, Z. (2015). Effects of high CO_2 treatment on green-ripening and peel senescence in banana and plantain fruits. *Journal of Integrative Agriculture, 14*, 875–887.

Sonmezdag, A. S., Kelebek, H., & Selli, S. (2014). Comparison of the aroma and some physicochemical properties of Grand Naine (*Musa acuminata*) banana as influenced by natural and ethylene-treated ripening. *Journal of Food Processing and Preservation, 38*, 2137–2145.

Srivastava, M. K., & Dwivedi, U. N. (2000a). Delayed ripening of banana fruit by salicylic acid. *Plant Science, 158*, 87–96.

Timilsina, U., Shreshta, A. K., Srivastava, A., & Shrestha, B. (2017). Ripening regulation and postharvest life improvement of banana MALBHOG using plant extracts and modified atmosphere package. *Asian Journal of Horticulture, 12*, 255–259.

Srivastava, M. K., & Dwivedi, U. N. (2000b). Delayed ripening of banana fruit by salicylic acid. *Plant Science, 158*, 87–96.

Stahler M. R. (1962). *Process for ripening bananas and citrus fruit*. U.S. Patent.

Stavroulakis, G., & Sfakiotakis, E. (1997). Regulation of propylene-induced ripening and ethylene biosynthesis by oxygen in 'Hayward' kiwifruit. *Postharvest Biology and Technology, 10*, 189–194.

Stepanova, A. N., & Alonso, J. M. (2005). Ethylene signalling and response pathway a unique signalling cascade with a multitude of inputs and outputs. *Physiologia Plantarum, 123*, 195–206.

Story, A., & Simons, D. H. (Eds.). (1999). *Fresh produce manual* (3rd ed.). Australian United Fresh Fruit & Vegetable Association Ltd.

Stover, R. H. (1972). *Banana, plantain and abaca diseases*. Slough: Commonwealth Agricultural Bureaux.

Stover, R. H., & Simmonds, N. W. (1987). *Bananas* (3rd ed.). London: Longmans.

Sultan, S., & Rangaraju, V. (2014). Changes in colour value of banana var. Grand Naine during ripening. *Bioscience Trends, 7*, 726–728.

Sun, L., Sun, Y., Zhang, M., Wang, L., Ren, J., & Cui, M. (2012). Suppression of 9-cis-epoxycarotenoid dioxygenase, which encodes a key enzyme in abscisic acid biosynthesis, alters fruit texture in transgenic tomato. *Plant Physiology, 158*, 283–298.

Supapvanich, S., & Promyou, S. (2013). Efficiency of salicylic acid application on postharvest perishable crops. In S. Hayat & A. A. M. Alyemei (Eds.), *Salicylic acid: Plant growth and development* (pp. 339–355). New York: Springer.

Surendranathan, K. K., & Nair, P. M. (1972). Properties of acidic and alkaline fructose 1,6- diphosphatease in gamma irradiated banana. *Phytochemistry, 11*, 119–123.

Surendranathan, K. K., & Nair, P. M. (1973). Alterations in carbohydrate metabolism of gamma irradiated Cavendish banana. *Phytochemistry, 12*, 241–246.

Surendranathan, K. K., & Nair, P. M. (1976). Stimulation of the glyoxalate shunt in gamma irradiated banana. *Phytochemistry, 15*, 371–774.

Surendranathan, K. K., & Nair, P. M. (1980). Carbohydrate metabolism in ripening banana and its alteration on gamma irradiation in relation to delay in ripening. *Journal of the Indian Institute of Science, 62*, 63–75.

Svanes, E., & Aronsson, A. K. S. (2013). Carbon footprint of a Cavendish banana supply chain. *The International Journal of Life Cycle Assessment, 18*, 1450–1464.

Tchango, J. T., Achard, R., & Ngalani, J. A. (1999). Etude des stades de recolte pour l'exportation par bateau, vers l'Europe, de trois cultivars de plantains produits au Cameroun. *Fruits, 54*, 215–224.

Thomas, P. (1986). Radiation preservation of foods of plant origin. III. Tropical fruits: bananas, mangoes, and papayas. *CRC Critical Reviews in Food Science and Nutrition, 23*, 147–206.

Thomas, P., & Janave, M. T. (1986). Isoelectric focusing evidence for banana isoenzymes with mono and diphenolase activity. *Journal of Food Science, 51*, 384–387.

Thomas, P., Dharkar, S. D., & Sreenivasan, A. (1971). Effect of gamma irradiation on the postharvest physiology of five banana varieties grown in India. *Food Science, 36*, 243–248.

Thompson, A. K. (2016). Postharvest transport. In *Horticulture compendium*. Oxford: CAB International.

Thompson, A. K., & Burden, O. J. (1995). Harvesting and fruit care. In S. Gowen (Ed.), *Bananas and plantains* (pp. 403–433). London: Chapman and Hall.

Thompson, A. K. (1985). Postharvest losses of bananas, onions and potatoes in PDR Yemen. *Tropical Development and Research Institute, London United Kingdom Contract Services Report CO 485.*

Thompson, A. K. (1996). *Postharvest technology of fruits and vegetables*. London: Blackwell Science.

Thompson, A. K., & Seymour, G. B. (1982). Comparative effects of acetylene and ethylene gas on initiation of banana ripening. *Annals of Applied Biology, 101*, 407–410.

Thompson, A. K., & Seymour, G. B. (1984). Inborja (CASCO) banana factory at Machala in Ecuador. *Tropical Development and Research Institute Report* (unpublished).

Thompson, A. K., & Silvis, H. (1974). Preliminary investigations on banana handling problems under Sudanese conditions. *Food and Agricultural Organization of the United Nations, Report Sud 70/543.*

Thompson, A. K., Been, B. O., & Perkins, C. (1972). Handling, storage and marketing of plantains. *Proceedings of the Tropical Region of the American Society of Horticultural Science, 16*, 205–212.

Thompson, A. K., Been, B. O., & Perkins, C. (1974). Effects of humidity on ripening of plantain bananas. *Experientia, 30*, 35–36.

Tiangco, E. L., Agillon, A. B., & Lizada, M. C. C. (1987). Modified atmosphere storage of 'Saba' bananas. *ASEAN Food Journal, 3*, 112–116.

Tadesse, T. N. (2014). Quality attributes and ripening period of banana (*Musa* spp.) fruit as affected by plant ethylene sources and packaging materials. *International Journal of Agricultural Research, 9*, 304–311.

Toan, V. N., Hoang, V. L., Tan, V. L., Thanh, D. C., & Luan, V. L. (2011). Effects of aminoethoxyvinylglycine (AVG) spraying time at preharvest stage to ethylene biosynthesis of Cavendish banana (*Musa AAA*). *Journal of Agricultural Science, 3*, 206–211.

Toan, V. N., Thanh, D. C., Le, V. H., Le, V. T., Truong, M. H., Thi Le, L. T., & Thi Thong, Q. A. (2010). Effect of near-harvest application of aminothoxyvinylglycine (AVG) on banana fruits during postharvest storage. *Acta Horticulturae, 875*, 163–168.

Toemmers, S., Blesgen, A., Kuhnen, F., Esdorn, L., Hass, V. C., & Hass, V. C. (2010). Model-based process control for optimised banana ripening. *International Federation of Automatic Control (IFAC) Proceedings, 43*, 317–322.

Toledo, T. T., Nogueira, S. B., Cordenunsi, B. R., Gozzo, F. C., Pilau, E. J., Lajolo, F. M., & Oliveira do Nascimento, J. R. (2012). Proteomic analysis of banana fruit reveals proteins that are differentially accumulated during ripening. *Postharvest Biology and Technology, 62*, 51–58.

Tongdee, S. C. (1988). *Banana postharvest handling improvements. Bangkok*: Report of the Thailand Institute of Science and Technology Research.

Tonutti, P. (2015). The technical evolution of CA storage protocols and the advancements in elucidating the fruit responses to low oxygen stress. *Acta Horticulturae, 1079*, 53–60.

Tressl, R., & Jennings, W. G. (1972). Production of volatile compounds in the ripening banana. *Journal of Agricultural and Food Chemistry, 20*, 189–192.

Truter, A. B., & Combrink, J. C. (1990). Controlled and modified atmosphere storage of bananas. *Acta Horticulturae, 275*, 631–638.

Tucker, G. A. (1993). Introduction. In G. Seymour, J. Taylor, & G. A. Tucker (Eds.), *Biochemistry of fruit ripening*. Cambridge: Cambridge University Press.

Turner, D. W. (1997). Bananas and plantains. In S. K. Mitra (Ed.), *Post harvest physiology and storage of tropical and subtropical fruits* (pp. 47–83). Oxford: CAB International.

Ullah, H., Ahmad, S., Anwar, R., & Thompson, A. K. (2006). Effect of high humidity and water on the quality and ripening of banana fruit. *International Journal of Biology, 8*, 828–831.

University of Queensland. (2017). file:///C:/Users/Anthony%20Keith%20Thomps/Downloads/UniQuest%20Tech%20Brief%20-%20Ripestuff%20Q3%202014.pdf. Accessed 24 March 2019.

USDA. (2012). *Nutrient database*. http://www.nal.usda.gov/fnic/foodcomp/Data/SR17/wtrank/sr17a306.pdf. Accessed 24 October 2012.

Vanderslice, J. T., Higgs, D. J., Hayes, J. M., & Block, G. (1990). Ascorbic acid and dehydroascorbic acid content of foods-as-eaten. *Journal of Food Composition and Analysis, 3*, 105–118.

Vendrell, M. (1969). Acceleration and delay of ripening in banana fruit tissue by gibberellic acid. *Australian Journal of Biological Sciences, 23*, 553–559.

Vendrell, M. (1970). Acceleration and delay of ripening in banana fruit tissue by gibberellic acid. *Australian Journal of Biological Sciences, 23*, 553–560.

Venkata Subbaiah, K., Jagadeesh, S. L., Thammaiah, N., & Chavan, M. L. (2013). Changes in physico-chemical and sensory characteristics of banana fruit cv. Grand Naine during ripening. *Karnataka Journal of Agricultural Sciences, 26*, 111 114.

Vermeir, S., Hertog, M. L. A. T. M., Vankerschaver, K., Swennen, R , Nicolai, B. M., & Lammertyn, J. (2009). Instrumental based flavour characterization of banana fruit. *LWT – Food Science and Technology, 42*, 1647–1653.

Vilela, C., Santos, S. A., Villaverde, J. J., Oliveira, L., Nunes, A., Cordeiro, N., Freire, C. S., & Silvestre, A. J. (2014). Lipophilic phytochemicals from banana fruits of several *Musa* species. *Food Chemistry, 162*, 247–252.

Von Loesecke, H. W. (1949). *Bananas*. London: Interscience.

Wade, N. L. (1974). Effects of O_2 concentration and Ethephon upon the respiration and ripening of banana fruits. *Journal of Experimental Botany, 25*, 955–964.

Wade, N. L. (1995). Membrane lipid composition and tissue leakage of pre- and early- climacteric banana fruit. *Postharvest Biology and Technology, 5*, 139–147.

Wade, N. L., O'Connell, P. B. H., & Brady, C. J. (1972). Content of RNA and protein of the ripening banana. *Phytochemistry, 11*, 975–979.

Wall, M. M. (2006). Ascorbic acid, vitamin A, and mineral composition of banana (*Musa* sp.) and papaya (*Carica papaya*) cultivars grown in Hawaii. *Journal of Food Composition and Analysis, 19*, 434–445.

Wang, C. Y., & Mellenthin, W. M. (1972). Internal ethylene levels during ripening and climacteric in Anjou pears. *Plant Physiology, 50*, 311–312.

Wang, Y., Luo, Z., & Du, R. (2015a). Nitric oxide delays chlorophyll degradation and enhances antioxidant activity in banana fruits after cold storage. *Acta Physiologiae Plantarum, 37*, 74. http://yadda.icm.edu.pl/yadda/element/bwmeta1.element.agro-c98ec1d2-ff21-4620-a29a-def5e3ea0093. Accessed 18 August 2019

Wang, Y., Luo, Z., Khan, Z. U., Mao, L., & Ying, T. (2015b). Effect of nitric oxide on energy metabolism in postharvest banana fruit in response to chilling stress. *Postharvest Biology and Technology, 108*, 21–27.

Wardlaw, C. W. (1937). Tropical fruits and vegetables: An account of their storage and transport. *Low Temperature Research Station, Trinidad Memoir 7,* Reprinted from Tropical Agriculture Trinidad, 14.

Wardlaw, C. W. (1961). *Banana diseases.* London: Longmans.

Wardlaw, C. W., & Leonard, E. R. (1940). *The respiration of bananas during ripening at tropical temperatures, studies in tropical fruits.* Low Temperature Research Station, Memoir 17.

Wardlaw, C. W., Leonard, E. R., & Barnell, H. R. (1939). *Metabolic and storage investigations of the banana.* Low Temperature Research Station, Memoir 11. Imperial College of Tropical Agriculture, Low Temperature Research Station.

Wei, Y., & Thompson A. K. (1993). Modified atmosphere packaging of diploid bananas (*Musa* AA). *Post-harvest Treatment of Fruit and Vegetables.* COST'94 Workshop, September 14 to 15 1993, Leuven.

Wendakoon, S. K., Ueda, Y., Imahori, Y., & Ishimaru, M. (2006). Effect of short-term anaerobic conditions on the production of volatiles, activity of alcohol acetyltransferase and other quality traits of ripened bananas. *Journal of the Science of Food and Agriculture, 86*, 1475–1480.

Wenkam, N. S. (1990). Food of Hawaii and the Pacific basin, fruits and fruit products: Raw, processed, and prepared. Volume 4: *Composition.* Hawaii Agricultural Experiment Station Research and Extension Series 110.

Wills, R. B. H., & Gibbons, S. L. (1998). Use of very low ethylene levels to extend the postharvest life of Hass avocado fruit. *International Journal of Food Properties, 1*, 71–76.

Wills, R. B. H. (1990). Postharvest technology of banana and papaya in Association of Southeast Asian Nations: An overview. *Association of Southeast Asian Nations Food Journal, 5*, 47–50.

Wills, R. B. H., & Tirmazi, S. I. H. (1982). Inhibition of ripening of avocados with calcium'. *Scientia Horticulturae, 16*, 323–330.

Wills, R. B. H., Harris, D. R., Spohr, L. J., & Golding, J. B. (2014). Reduction of energy usage during storage and transport of bananas by management of exogenous ethylene levels. *Postharvest Biology and Technology, 89*, 7–10.

Wills, R. B. H., Ku, V. V. V., & Leshem, Y. Y. (2000). Fumigation with nitricoxide to extend the postharvest life of strawberries. *Postharvest Biology and Technology, 18*, 75–79.

Wills, R. B. H., McGlasson, B., Graham, D., & Joyce, D. (1998). *Postharvest: An introduction to the physiology and handling of fruit, vegetables and ornamentals* (4th ed.). Wallingford: CAB. International.

Wills, R. B. H., Poi, A., Greenfield, H., & Rigney, C. J. (1984). Postharvest changes in fruit composition of *Annona atemoya* during ripening and effects of storage temperature on ripening. *HortScience, 19*, 96–97.

Wilson Wijeratnum, R. S., Jayatilake, S., Hewage, S. K., Perera, L. R., Paranerupasingham, S., & Peiris, C. N. (1993). Determination of maturity indices for Sri Lankan Embul bananas. *Australian Centre for International Agricultural Research Proceedings, 50*, 338–340.

Woolf, A. B., Wexler, A., Prusky, D., Kobiler, E., & Lurie, S. (2000). Direct sunlight influences postharvest temperature responses and ripening of five avocado cultivars. *Journal of the American Society for Horticultural Science, 125*, 370–376.

Wu, B., Guo, Q., Li, Q., Ha, Y., Li, X., & Chen, W. (2014). Impact of postharvest nitric oxide treatment on antioxidant enzymes and related genes in banana fruit in response to chilling tolerance. *Postharvest Biology and Technology, 92*, 157–163.

Yang, S. F. (1981). Biosynthesis of ethylene and its regulation. In J. Friend & M. J. C. Rhodes (Eds.), *Recent advances in the biochemistry of fruit and vegetables* (pp. 89–106). London: Academic.

Yang, S. F., & Ho, H. K. (1958). Biochemical studies on post-ripening of banana. *Journal of the Chinese Chemical Society, 5*, 1–98.

Yang, S. F., & Hoffman, N. E. (1984). Ethylene biosynthesis and its regulation in higher plants. *Annual Review of Plant Physiology, 35*, 155–189.

Yang, S. F., & Pratt, H. K. (1978). The physiology of ethylene in wounded plant tissue. In G. Kahl (Ed.), *Biochemistry of wounded plant tissues* (pp. 595–622). Berlin: Walter de Gruyter.

Yang, S. F., Adams, D. O., Lizada, C., Yu, Y., Bradford, K. J., Cameron, A. C., & Hoffman, N. E. (1979). Mechanism and regulation of ethylene biosynthesis. In F. Skoog (Ed.), *Plant growth substances 1979* (Proceedings in Life Sciences) (pp. 219–229). Berlin/Heidelberg: Springer.

Yao, A. K., Koffi, D. M., Irié, Z. B., & Niamke, S. L. (2014). Conservation de la banane plantain (*Musa* AAB) à l'état vert par l'utilisation de films de polyéthylène de différentes épaisseurs. *Journal of Animal & Plant Sciences, 23*, 3677–3690.

Yoo, S.-D., Cho, Y., & Sheen, J. (2009). Emerging connections in the ethylene signaling network. *Trends in Plant Science, 14*, 270–279.

Youryon, P., & Supapvanich, S. (2017). Physicochemical quality and antioxidant changes in 'Leb Mue Nang' banana fruit during ripening. *Agriculture and Natural Resources, 51*, 47–52.

Zaman, W., Paul, D., Alam, K., Ibrahim, M., & Hassan, P. (2007). Shelf-life extension of banana (*Musa sapientum*) by gamma radiation. *Journal of Bio-Science, 15*, 47–53.

Zhang, M., Jiang, Y., Jiang, W., & Liu, X. (2006). Regulation of ethylene synthesis of harvested banana fruit by 1-MCP. *Food Technology and Biotechnology, 44*, 111–115.

Zhang, M., Leng, P., Zhang, G., & Li, X. (2009). Cloning and functional analysis of 9-cis-epoxycarotenoid dioxygenase (NCED) genes encoding a key enzyme during abscisic acid biosynthesis from peach and grape fruits. *Journal of Plant Physiology, 166*, 1241–1252.

Zhang, M., Yuan, B., & Leng, P. (2009a). Cloning 9-cis-epoxycarotenoid dioxygenase (NCED) genes and the role of ABA on fruit ripening. *Plant Signalling and Behaviour, 4*, 460–463.

Zheng, Y., Fung, R. W. M., Wang, S. Y., & Wang, C. Y. (2008). Transcript levels of antioxidative genes and oxygen radical scavenging enzyme activities in chilled zucchini squash in response to super-atmospheric oxygen. *Postharvest Biology and Technology, 47*, 151–158.

Zhu, X., Luo, J., JunLi, Q. L., Liu, T., Wang, R., Chen, W., & Li, X. (2018). Low temperature storage reduces aroma-related volatiles production during shelf-life of banana fruit mainly by regulating key genes involved in volatile biosynthetic pathways. *Postharvest Biology and Technology, 146*, 68–78.

Index

Printed in the United States
By Bookmasters